Man, A Geomorphological Agent

MAN, A GEOMORPHOLOGICAL AGENT

An Introduction to Anthropic Geomorphology

DOV NIR
The Hebrew University, Jerusalem

Keter Publishing House
Jerusalem, Israel

D. Reidel Publishing Company
Dordrecht : Holland / Boston : U.S.A. / London : England

Library of Congress Cataloging in Publication Data

CIP

Nir, Dov.
Man, a geomorphological agent.

Bibliography: P.
Includes index.
1. Man — Influence on nature. 2. Geomorphology.
I. Title.
GF75.N.56 304.2 82-6558
ISBN 90-277-1401-0 (D. Reidel) AACR2

Published by Keter Publishing House, Israel,
in co-publication with D. Reidel Publishing Company,
P.O. Box 17, 3300 AA Dordrecht, Holland.

Sold and distributed in the U.S.A. and Canada
by Kluwer Boston Inc.,
190 Old Derby Street, Hingham, MA 02043, U.S.A.

Sold and distributed in Israel
by Keter Publishing House,
P.O. Box 7145, Jerusalem, Givat Shaul B, Israel.

In all other countries, sold and distributed
by Kluwer Academic Publishers Group,
P.O. Box 322, 3300 AH Dordrecht, Holland.

D. Reidel Publishing Company is a member of the Kluwer Group.

Edited by David Ashboren

Printed in Israel by Keterpress Enterprises, Jerusalem

To Ruth
and Vered, Moshe, Zili, Gideon
and Guy

Man must be regarded directly as a geomorphological agent, for he has increasingly altered the conditions of denudation and aggradation of the Earth's surface.

CARL O. SAUER (1931)

Contents

Foreword

This book intends to study the nature and rate of Man's activity in sculpturing (and degrading) the landscape. It is not just another book on environmental geomorphology dealing with the influence of the geomorphological processes on Man: its standpoint regards Man as an independent geomorphological agent, whose activities are guided by his social, economic, and iconographic needs and opportunities. If today the amount of slags from the Witwatersrand mines is growing at a daily rate of 165,000 tons; if in the open pit mine of Bingham Canyon Copper Mine in Utah the volume of rock produced daily is some 330,000 tons (McWhirter and McWhirter, 1975), then the need for a study such as this book seems quite obvious. The subtitle of the book—An introduction to anthropic geomorphology—suggests only one aspect of the subject matter. An second subtitle—A chapter in cultural geography—would fit also its subject matter, only considered from a different point of view.

The organization of the book follows this concept: its chapters are not following a scheme of geomorphological phenomena, but center around sectors of human activity. The introduction deals with the background of the problem—the basic laws of geomorphological processes, superimposed on an ever-growing population having ever-expanding possibilities for physical interference with the landscape. Following that, the first chapters focus on the use of the various resources: grazing, forest, working the soil, and mining. The following chapters deal with Man's transportation means, river and sea shore management, and, finally, with the influence of built-up areas on geomorphological processes. In conclusion an attempt is made to evaluate the rate of the geomorphological processes induced by Man. I hope that this book will elucidate the specific position anthropic geomorphology deserves in geography.

In an integrative study of this type, authorities in geography, geology, biology, engineering, and history have had to be consulted. I enjoyed and appreciated the aid and hospitality of the School of Geography of the University of Oxford and of its

Head, Professor Jean Gottmann, during my Sabbatical leave from the Hebrew University of Jerusalem. Miss Buxton, the Librarian of the School, and her assistants gave me invaluable guidance through the Oxonian system of the libraries. I would also like to thank the librarians of other libraries which I consulted. These include the Ratcliff Science Library; the Oxford Department of Geology; the Commonwealth Institute of Forestry; the Rhodes Institute; the Oxford Departments of Agricultural Sciences and of Engineering Sciences; the Map and Geography Room at the Bodleian Library; and, finally, the British National Library at Whetherby. I want to thank the following for the use of illustrative materials: Fig. 1, Strahler, 1956; Fig. 2, Darby, 1956; Fig. 3, Eaux & Forêts, Gov. of France; Fig. 4, Curtis, 1956; Figs. 5 and 6, Conacher and Murray, 1973; Fig. 8, Bishop and Stevens, 1964; Fig. 9, Mensching, 1951; Fig. 10, Stocking, 1972; Fig. 11, Curtis, 1972; Fig. 12, Haserodt, 1971; Fig. 13, Spencer and Hale, 1961; Fig. 14, Hempel, 1971; Fig. 15, Mead, 1954; Figs. 16 and 17, Smith, 1975; Fig. 18, Aufrère, 1929; Fig. 22, Whitaker, 1971; Fig. 25, Denevan, 1970; Figs. 26 and 27, Parsons and Bowen, 1966; Fig. 28, Tóth and Fekete, 1974; Fig. 29, Nat. Geogr. Mag., 1964; Fig. 30, Thomas, 1956; Fig. 31, U.K. Ministry of Town and Country Planning, 1953; Fig. 34, Parizek, 1971; Fig. 38, F. H. Taylor, 1938; Fig. 40, Chi-sen Liang, 1973; Fig. 41, U.K. Department of Environment, 1972; Fig. 43, Daniels, 1960; Fig. 44, Ruhe, 1971; Fig. 45, Collier, 1972; Fig. 47, Kassas, 1972; Fig. 48, Meade and Trimble, 1974; Fig. 49, Royal Commission on Coast Erosion and Afforestation, U.K., 1911; Fig. 51, Netherland's Government Information Service, 1967; Fig. 52, Chi-sen Liang, 1973; Figs. 53 and 54, Harris and Rantz, 1964; Fig. 55, Guy and Ferguson, 1970; Fig. 56, Maderey, 1974; Fig. 57, Chen, 1974; Figs. 58, 60 and 61, Lindh, 1972; Fig. 59, Seaburn, 1969. All other illustrations are originals and appear here for the first time; the drawings have been produced by Mr. Naftali Baer, to whom I am very grateful. My deep appreciation is due to the staff of the Keter Publishers, who edited and carried out the publication of this book. Last but not least, I would like to thank Dr. David Ashboren who edited the manuscript and gave it its final shape.

Dov Nir

Jerusalem, September 1981

Introduction

THE PERCEPTION OF MAN'S GEOMORPHOLOGICAL ACTIVITY

A scholar reviewing the geomorphological literature of the last 40 years will gain the impression that the perception of Man as a geomorphological agent is a fairly recent development, stemming from from the much earlier concept of environmental geomorphology. The important treatises and textbooks of geomorphology have paid scant attention to this aspect of geomorphological processes. E. de Martonne's *Traité de géographie physique* makes no mention of human involvement in physical geography; the early editions of *Précis de géomorphologie* by M. Derruau do not deal with anthropic geomorphology, while in the fifth edition (1967) only two and one-half pages are devoted to this subject, but even these pages are essentially a review of soil erosion accelerated by Man, no coverage being given to urban, mining, or other human activities. Doerr and Guernsey (1956) complain of the lack of perception of Man's geomorphological activity in scientific circles:

> A random inventory of geology, geography, physical geography and physical geology reveals that most of the authors ignore Man as a geomorphological agent: E. de Martonne, C. A. Cotton, F. Machatschek, O. D. von Engeln, A. K. Lobeck, V. C. Finch, G. T. Trewartha, H. Baulig.

Such an extreme statement is, of course, not entirely valid. Louis, in his *Allgemeine Geomorphologie* (1960), deals very effectively (pp. 272–279) with this aspect. However, the perception of Man's role in geomorphology, which seemingly has come to the fore only in the late 1960s, actually had very noteworthy predecessors. It was only due to negligence and the prevailing geomorphological trends and ideas during successive periods that the forerunners of anthropic geomorphology were left in the shadow. Nevertheless, the concept did exist, if in an embryonic state.

From Marsh to Sherlock

In 1864, G. P. Marsh made the following drastic statement *(Man and Nature, or Physical Geography as Modified by Human Action)*:

> Man is everywhere a disturbing agent. Wherever he plants his foot, the harmonies of Nature are turned to discord.

Marsh was the first of the modern scholars to recognize Man as a factor, and not merely as a subject, in Nature. This was at a time when the deterministic school, influenced by Darwin's theory of evolution, was the ruling one in the natural sciences and even in geography. However, Marsh's prime interest was the biological impact of Man's influence, and the purely geomorphological aspect was secondary to him. Nevertheless, he made a very important contribution to the comprehension of the destructive influence of forest clearing, which triggers slope erosion, sedimentation in the river beds, and, subsequently, disastrous floods. Marsh himself admits that he was certainly not the first to gain awareness of the damage caused by the cutting of forests (*loc. cit.*, p. 188):

> The knowledge of the evil of the influence of the cutting of forest is not new; even Pliny knew it: "Destruction torrents are generally formed when hills are stripped of the trees which formerly confined and absorbed the rains."

Marsh traveled extensively outside the U.S.A., mainly in the Mediterranean region, and perhaps his encounter with the Mediterranean landscapes inspired him to state (*loc. cit.*, p. 47):

> ... destructive changes are most frequent in countries of irregular and mountainous surface, and in climates where the precipitation is confined chiefly to a single season, and where the year is divided into a wet and dry period, as it is in the Mediterranean basin.

Thus Marsh introduced the importance of the *steepness of slopes* and *rainfall intensity* as coefficients by which the human intervention in natural processes should be multiplied. However, both the European and the American geomorphological literature in the last third of the 19th century lack even allusions to Marsh's view on the role of human activity in natural geomorphological processes. Perhaps the geographical and geological discoveries in the Western U.S.A., Africa, Asia, and elsewhere overshadowed the other aspects of geomorphology, as did the efforts to elaborate a geomorphological model the climax of which was the theory of W.M. Davis on the geographical cycle (1899).

It was only in 1901 that a paper by Woeikof ("De l'influence de l'homme sur la nature") followed Marsh's earlier trend. The main issue dealt with by Woeikof is the felling of forests and the gullying introduced by the destruction of the natural vegetation, as well as by drainage and irrigation. He lays the blame on the rapid development of metropolises and the growing need for food and raw materials, for the hasty extension of fields at the expense of forests. The evil is the "dissociation between man and the soil," and the agglomeration of the population in the town, which in itself is evidence of an unhealthy state. With this evaluation Woeikof became the first scholar to introduce social and iconographical values into anthropic geomorphology, his opinions and thoughts being a reflection of the Tolstoyan philosophy popular at the end of the 19th and the beginning of the 20th centuries.*

The concept of geomorphological processes as an integral part of the human environment, and as a domain of active human intervention, also existed in the United Kingdom, but there it took a surprisingly practical turn. In 1906 a Royal Commission was appointed, its task being (Royal Commission on Coast Erosion and Afforestation, 1911)

> . . . to inquire and report as to the encroachment of the sea on various parts of the coast of the United Kingdom; and the damage which has been or is likely to be caused hereby; and *what measures are desirable for the prevention of such damage.*

It is not by chance that investigation on environmental geomorphology in the U.K. seemed to be the most advanced for its time; by the turn of the century the most effective shaping of the landscape by Man's hand was evident in the British Isles. Whereas in most of the countries contemporary human intervention mainly took the form of grazing, foresting, and agricultural activities, the landscape of the United Kingdom was sculptured by open mining, railway and highway cuttings, the drainage of fens, and the most urbanized society of the end of the 19th century (England's scenery, *Encyclopaedia Britannica*, Vol. 6, p. 867, 1974):

> Only the remotest moorland and mountain tops have been untouched; there is virtually no genuine wilderness left in England.

The perception of anthropic geomorphology in this period culminated in the paper by Fischer (1915) and in Sherlock's book (1922), the most important work ever written on the subject.

Fischer, in his paper "Der Mensch als geologischer Faktor" reprimands contemporary geologists for not showing interest even in young alluvial deposits, not to speak of human interference with Nature. To the best of our knowledge, he made the first estimate of the volume of minerals utilized by man, and also of the amount of gravel, sand, and stone quarried for building purposes. He deals with waterways, drainage, swamp reclamation, and forest management.

In Sherlock's *Man as a Geological Agent* . . ., the diverse human activities are brought into question for the first time. Not only forest, grazing, and agricultural management but also the various aspects of the technological and industrial activity of a modern, 20th-century society are evaluated in terms of their intervention in natural processes: construction of highways and railways; open pit mining of coal, slate, stone, gravel, and sand; waterways and coastal management.

Sherlock managed to provide statistics for the volumes of material extracted or removed, and he gives an estimate of the rates of erosion and sedimentation. He deals mainly with the United Kingdom, since most of the statistics available to him related to that country. However, the importance of the book is universal. This is the first serious treatise with a quantitative approach to the problem of human geomorpho-

* Perhaps the present-day interest in the environment is a return to this kind of philosophy.

logical activity. It is surprising that this work has not been cited in most of the geomorphological textbooks which have appeared in the 40 years since its publication.

From Sherlock to Brown

The late 1920s and the 1930s were dominated by the recognition of Man's deleterious influence on the soil and the need for *soil conservation*, an idea propagated mainly in the U.S.A. by Bennett (1928, 1939, 1955). This new awareness has had far-reaching effects throughout the globe, and it has led to better soil management and to improved agricultural techniques. But because of its domination, the main trends of the galloping urbanization were not yet recognized as a factor shaping the Earth; this perception of urban geomorphology had to wait until the 1960s before it gained acceptance by the geologists and geomorphologists, mainly in the U.S.A.

However, anthropic geomorphology was not generally a topic of research; here and there scholars touched upon it. Aufrère (1929), in his work on the agricultural ridges ("rideaux") of northern France, entitled one of his chapters "L'Homme, agent géomorphologique." Fels (1935) dealt with geomorphology in his *Der Mensch als Gestalter der Erde.*

Anthropic geomorphology reemerged after World War II. At that time an attempt was being made to provide geomorphology with a scientific basis by laying emphasis on quantification and by analyzing the basic processes. In 1951, Mensching correlated the deposition of silt in the Weser valley with the historic deforestation of its upper reaches; in 1953, Tricart employed the term "homme, agent d'érosion," and attributed to Man the introduction into Nature of mechanical activities hitherto unknown.

Man's Role in Changing of the Face of the Earth, the proceedings of a symposium published in 1956, should be considered a landmark in anthropic geomorphology. It heralded renewed interest in the subject, as emerges from the work of Legget (1962):

> One tenth of the land surface of England and Wales is covered with buildings, railways, roads, etc.; in the U.S., the roads and streets only are covering one percent of the total land area.

A comprehensive analysis was provided by Golomb and Eder (1964). But even in 1965, there were complaints about the inadequate role apportioned to anthropic geomorphology in geomorphological research (Smith and Bridges, 1965):

> All these examples — burning of vegetation, the effects of the 18th- and 19th-century metallurgical industry, activities of sheep — are accountable to the work of Man, but the widespread nature of such acting seems to stress *how meagre is our information on this topic.*

Anthropic geomorphology had been given a new impetus. In 1960, Zapletal classified changed landscapes into categories, varying from the unchanged, natural to the completely cultural ones. Researchers working in such remote areas as Poste de la Baleine in Quebec, Canada, perceived that even the primitive activity of Man induced geomorphological changes (Cailleux and Hamelin, 1969). An important contribution was made by Jennings (1965). The mid-1960s saw the rise of a new wave of research into hitherto neglected topics: the influence of urban development on the natural environment. Of the many authors dealing with this new, rapidly developing branch of geomorphological research we shall mention only Diseker and Richardson (1961, 1962), Savini and Kammerer (1961), Keller (1962), Harris and Rantz (1964), Wolman (1967), Wolman and Schick (1967), Leopold (1968), Martens (1968), Seaburn (1969), McPherson (1969), and Brater and Sangal (1969). This period of reawakening and rediscovery has been summed up by Brown ("Man Shapes the Earth," 1970).* He methodically distinguishes between direct, purposeful human action, direct but incidental human action, and the involuntary results of Man's modification of geomorphological processes.

Such direct actions are the felling of trees, agricultural techniques, such as ridging and terracing, the removal of steep slopes for housing developments, filling of gaps for roads, etc. Examples of direct but incidental activities are the pits left after the extraction of useful material, such as gravel or sand. However, amazing as it may seem, by far the most important activities from the quantitative aspect are those which, by themselves, were not intended at all: the sealing of surfaces by concrete or macadam; slope erosion following ploughing and grazing; coastal processes induced by jetties, groynes, etc.

As regards the spread of anthropogeomorphological activities, Brown sees their ever-growing weight resulting from the concentration of Man's activities on small areas, from the increased efficacy of the tools effecting the natural environments, and from the increasing population of the world.

In 1968, the U.S. National Academy of Sciences appointed a special commission on geological sciences to study "the new developments in and potential application of the geological sciences." In 1972 this commission published a volume under the editorship of Collier, entitled *The Earth and Human Affairs*, a general review of anthropic geomorphology. Strahler and Strahler (1973) provide a 27-page discussion *(Environmental Geosciences: Interactions Between Natural Systems and Man)* of the direct human influence on natural processes. H. F. Game, in his *Origins of Landscapes* (1974), devotes a whole chapter (pp. 677–693) to environmental geomorphology, stating that "if this chapter is the last and the shortest, it is not the least important." More recent works are those of Eckholm (1976), Howard and Remson (1978), and Rathjens (1979). Since 1978 anthropic geomorphology has come into its own as a fully accepted branch of geomorphology (Haigh, 1978).

DEMOGRAPHIC, SOCIAL, AND ECONOMIC FACTORS IN ANTHROPIC GEOMORPHOLOGY

Man's intervention in the natural landscape began with his first steps; like all the creatures, his mere existence constitutes an integral part of the processes shaping the environment. However, in contrast to all other creatures, man possesses *tools* which enable him to intensify his intervention. His tools are a direct result of his tech-

* Inaugural lecture at University College, London, February 1968.

nological and economic advancement (Bennett, 1939):

> Since the first crude plow uprooted the first square foot of sod, and since Man's axe first bit in the virgin forest, erosion of soil has been a problem. It is as old as history.

It is quite inadequate to view Man's efforts as relating solely to his adaptation to nature; beginning by picking up any broken piece of stone useful to his purpose, he has progressed in adapting surrounding nature to his needs (Sherlock, 1922).

From the very outset, this intervention has had two aspects: biological and geomorphological. Man interfered in *biological* processes already in the most early period (Tricart, 1953); destruction of plants and animals by fire; gathering, hunting, and grazing. His later, more sophisticated intervention included breeding, the creation of new species of plants and animals (dogs, sheep), and the transfer of plants from their indigenous environment to faraway, different surroundings (cereals, fruits). Instead of the original biological environment, a new one was created; often this new environment was at variance with the dominant climatic or topographical conditions of the area.

The *geomorphological* aspects are, as already mentioned, those involving the destruction of the soil cover and the subsequent denudation and erosion. The intensity of Man's intervention differed from one country to another, and from one historical period to the next, depending on the type of society. Irrigation which changed the courses of rivers, drainage of marshes, swamps, and even parts of the sea, ploughing, improvement of slopes by terracing, and much more: all these activities were geomorphological agents of prime importance but were rarely noted by man as such except in cases of accidents or disaster.

The activity of a single man cannot alter a landscape very much; however, Man does not act as an individual but as part of a society. Therefore, anthropic geomorphology is greatly dependent on the social, i.e., demographic, economic, and iconographic aspects of humanity. The research of anthropic geomorphology, evaluating the intensity, volume, and rate of human intervention in the environment, should therefore take into account the following considerations:

1. The destruction by a single human activity and its global impact on the Earth's surface (Collier, 1972), by multiplying it by the world population, i.e., the *demographic* factor.

2. The time interval over which these activities have occurred (intensive land occupancy in some areas has continued without a break for 6,000 to 15,000 years (*ibid.*, 1972), i.e., the *historical* factor.

3. Man's increasing capacity to modify the landscape by new technologies and by capital investment, i.e., the *economic* factor.

4. The growing demand and need for a higher standard of living, which involves further intervention in natural processes, i.e., the *socioeconomic* factor.

Even though the anthropogeomorphological processes have existed since the arrival of Man on Earth, their present powerful impact results from their *omnipresent dispersion and their unlimited scale*; their influence is mainly, if not entirely, a result of the quantities (of earth, rock, water, etc.) involved. The quantity of products depends, first and foremost, on the *population density* in any area. This density is steadily increasing almost everywhere in the world, a phenomenon well known to the reader and requiring no elaboration here.

Let us, for example, briefly review the history of human activity and its impact on geomorphology in Wales. This country, like most other parts of the globe, was very sparsely populated in Neolithic times, with less than one person per square kilometer (Fairbrother, 1972). During the Iron Age the estimated population density was four persons per square kilometer; in the Middle Ages it went up to 50, reaching 318 in 1970. These average figures do not reveal greater concentrations occurring in certain areas. Even if we do not take into account expanding economic and technological activity and attribute to the population the same degree of activity in all periods, the total activity must have increased more than 300 times since the Neolithic period, some 80 times since the Iron Age, and more than six times since the Middle Ages. If we were to elaborate a coefficient of technological and economic advancement (plowing, building, damming, mining, macadamizing, etc.), we would note the immense increase in the volume of anthropogeomorphological processes wrought by the changing human society living in Wales. Not all countries have undergone technological evolution and experienced a population increase to the same degree as Wales. However, a general increase in population, and particularly a steep rise in its density in certain localities called "cities," is a worldwide phenomenon. If we regard the population density as the basic parameter of anthropic geomorphology, its effects should be rather small at the beginning of the Christian era; the estimated world population at that time was in the order of 200 million (Weitz, 1973). The population is reckoned to have grown very slowly, and by 1600 stood at some 400 million. In 1750 it had expanded to 694 million, giving over the preceding century an annual rate of increase of 0.4%. In the following century, the annual rate of growth was similar—0.45%, with 1.091 millions by 1850. This rate almost doubled in the next 50 years (0.70% per year), and the world's population reached 1,550 million by 1900. The growth in the first half of the 20th century was even greater; the annual rate was twice that in the previous 50 years—1.30%, and the population reached 2,454 million in 1950. In the subsequent decade the total population increased by the same absolute number as in the previous quarter of a century, and stood at 2,982 million—an annual increase of more than 2%. In 1976 the world's population stood at 4,044 million (*United Nations Statistical Yearbook*, 1977), having grown over 16 years at an annual rate of 2.2%; in 1980, the world's population attained 4.2 milliard. This last increase since 1960 by itself exceeds almost the entire population of the globe in 1850.

The population explosion of this century should be viewed in terms of the needs for natural raw materials for construction, such as timber, stone, gravel, and sand, for roads, for water, and for ploughing and grazing, all of which have rapidly increased in the last 100 years. All these activities have been stimulated by the new technology, which can move earth material in quantities and qualities which were unthinkable in the previous generations. Tractors, for example, which were unknown at the beginning of the century, numbered in 1976 some 18 million units (*United Nations Statistical Yearbook*, 1977). The Gross National Product (GNP), an indicator of expanding economic capacity, grew in every country. This last decade was marked by greater demands for better housing, wider highways, and diverse activities for leisure time (management of beaches for swimming, slopes for skiing, rivers for yachting).

All these social, technological, and demographical trends govern the intensity of the geomorphological development. We cannot deal successfully with the latter without taking into account the former; in fact, every human activity on (and below) the Earth's surface causes a reaction in the natural environment. *Mankind is part of a system* in which changes in the geomorphology play a dominant role. Our task is to evaluate his role in the system.

THE CONCEPT OF "STEADY STATE" IN GEOMORPHOLOGY

The Landscape as a System

It is difficult to treat separately the vegetation, the topography, the movement of water, and the sedimentation of soil in a given landscape; they are closely related and interlinked, constituting a single system. A system is characterized by the interaction of its parts. Every change in the nature, or even in the intensity of a particular process, has an effect on the system, and under critical or limiting conditions may radically alter the entire system or even destroy it. The size of a system is not a criterion for assessing the intensity of the interactions occurring inside it. The world, of course, is a system of systems, but a clearly defined and limited slope is also a system, as is a lake, a river bed, or an oasis.

In order to study the geomorphological factors operating in a system, one needs to know how the different biological, climatological, and topographical factors function within it. Every change in these factors will have an impact, the intensity of which depends on the geomorphology of the system. This statement seems so obvious that there should be no need of even mentioning it here. However, the crucial point is that this "obvious" knowledge is widely neglected by active persons dealing with their environment, causing the well-known *negative* anthropic geomorphological effects: erosion, wasting of landscapes, and even disasters.

The basic factors affecting landscape are climate, soil, topography, plant cover, and running water (Smith and Wischmeier, 1962). Of these, climate is beyond the sphere of Man's activity. Even though it is possible to introduce some important modifications in the topoclimatic elements, the basic meteorological conditions, such as the movement of air masses, are beyond human control. Three of the factors—topography, plant cover, running water—are within the limits of Man's activity. While soils cannot be created entirely by Man, they can at least be modified for better or for worse.

There is no general scale for measuring human intervention in the landscape system; certain zones are affected to a greater, others to a lesser extent. The geomorphological effect of cutting a highway in a loess region, creating slopes of 25° to 30°, will not be the same in a humid, temperate zone as in a semiarid, hot zone.

Whereas in the former zone no special steps need be taken to stabilize the slope, since it will be colonized by weeds within a short span of time, in the latter zone this same cutting, if not protected by introduced vegetation, will trigger off heavy soil erosion. In both cases, the highway cutting called for the same degree of human intervention, but different environmental systems were involved. Is this situation being taught at the technical universities?

The "Steady State" in Geomorphology

A system is maintained in its form by the equilibrium existing between the different processes and forces operating. Chorley (1962) defined geomorphological processes as parts of an *open system* which has an input of certain energy and operative forces (rain, wind, force of gravity), performs work in a certain area (watershed), and produces a new landscape (erosion and sedimentation) as output. All parts of this system are governed by an equilibrium between the energy input and the resulting output (the actual landscape).

This modern notion of landscape equilibrium was anticipated more than 100 years ago by Marsh (1864, ed. 1965, p. 29):

> Nature, left undisturbed, so fashions her territory as to give it almost unchanging permanence of forms, outline and proportions, except when shattered by geological* convulsions; and in these comparatively rare cases of derangements, she sets herself at once to repair the superficial damage and to restore, as nearly as practicable, the forms as part of her domination. ...
>
> By a general compensation of conflicting forces, a *condition of equilibrium* has been reached which, without the action of man,** would remain, with little fluctuation, for countless ages.

It appears that Marsh, when commenting on "landscape," mainly had in mind a biological equilibrium (forests, animals) rather than a geomorphological one. However, the notion of equilibrium has also been adopted into geomorphology, but in a peculiar form—the "steady state." This notion of steady state differs from the more general and perhaps misunderstood term "equilibrium" in that it does not allude to inactivity, but to a *stability of active processes.* The concept of steady state does not imply that the processes are not operating; on the contrary, it points out that, because of a degree of relationship between the active forces, the elements active in the system are in equilibrium with one another at a certain time and in a certain place. This steady state is related to the entropy of the acting forces (Chorley, 1962); the forces utilize a minimum of energy merely to maintain the steady state. Let us elucidate this pattern by examples of steady state existing in a stream and on a slope.

The concept of equilibrium can be useful in dealing with the response of channel systems to significant changes in discharge and sedimentation (Wolman, 1967, 1974). These are related to the soil, lithology, vegetation, topography, and climate of the region. Both the form of a river channel and the deposits associated with it at any

* We should add: and climatic.

** We should add: and catastrophic events of geological (and climatic) origin.

given time and place represent the interaction of basic processes of fluid flow and sediment transportation. Any change in the fluid flow affects the equilibrium between the channel form and its deposits and vice versa. The basic rate of flow (assuming plug flow) can be expressed by the simple equation

$$Q = A \cdot V$$

where the discharge Q is the volume rate of flow; A is the area of cross section of the flow; and V is the velocity of the flow. For the sake of simplification, the area A can be expressed as the product of the width and the depth of the channel, provided that its shape is fairly rectangular.

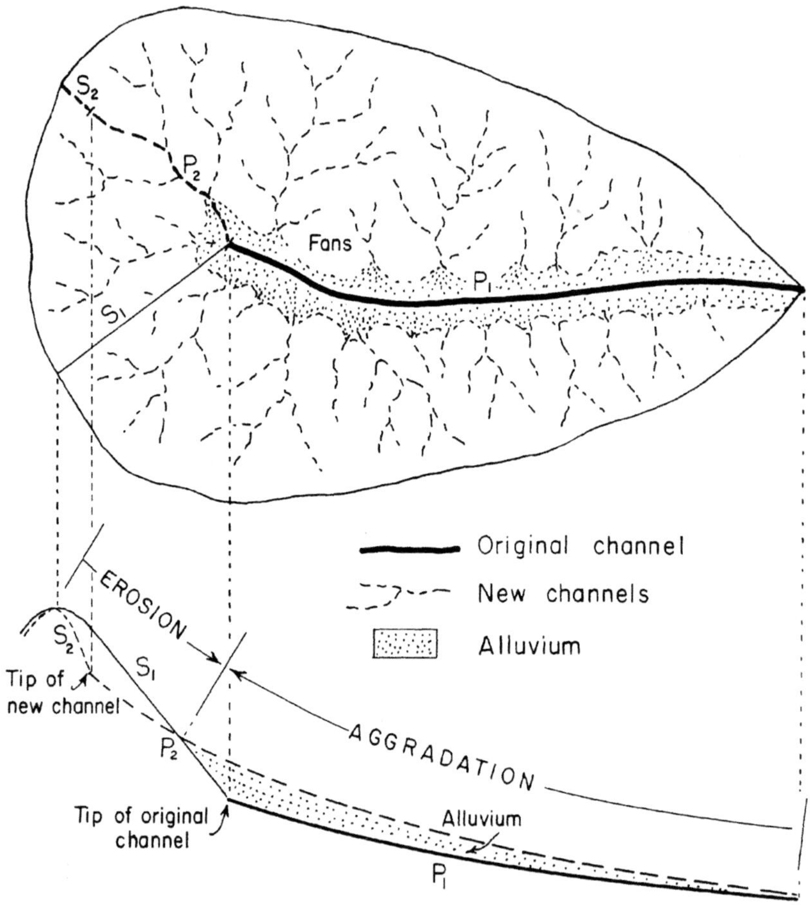

FIG. 1. Drainage-basin transformation from low- to high-drainage density, as a result of a breakdown of initial surface resistivity and reduction of infiltration capacity (Strahler, 1956)

Any change in one of these basic variables, Q, A, or V, will induce a change in the equilibrium and find expression in the shape of the channel and the processes occurring in it. Deepening or silting up of the channel, a change in the bottom slope (inducing a change in velocity), and a decrease or increase in the volume of the flow, all these will influence the *form* of the channel itself by erosion or deposition which did not occur in the steady state (as expressed by the equation) prior to the change.

This steady-state model can also be applied to other elements of the landscape, such as slopes. A slope is part of a landscape having a certain surface and inclination, which is in equilibrium between the force of gravity and the forces of friction acting between the particles constituting the slope. Any change in the steady state existing between these forces without a parallel change in the slope, such as the removal of vegetation, triggers the activity of gravity, which is no longer in equilibrium with the friction forces, and a movement of the particles forming the surface of the slope can be produced (Fig. 1). This movement, an adaptation to the new interaction of forces, can lead to accelerated erosion, a gully system developing on a previously smooth unchanneled slope (Strahler, 1956). A change in a related biological factor (vegetation) greatly enhances a geomorphological factor (erosion). For a given runoff, more material will be eroded from a bare surface than from a covered one.

The land formations on which we live are only apparently stable; actually, they are in a steady state, resulting from a temporary equilibrium among the forces active in the landscape. This steady state is very precarious and fragile. If left alone, this equilibrium can be quasi-permanent, except for catastrophic natural events. There is, however, virtually no spot on the globe (with the exception, perhaps, of remote deserts and ice sheets) that is in a truly "natural" state.

Man's Intervention in the Steady State

Man's intervention in Nature is defined as a disturbance of the existing natural equilibrium. Such disturbance is brought about in various ways.

Man's intervention against the force of gravity

The force of gravity is the main moving force in geomorphological processes. Aside from internal forces (volcanism, earthquakes) and the dynamics of the wind, it is the force of gravity which naturally moves mases of water in channels, or detritus. However, Man can act against the natural force of gravity (Strahler and Strahler, 1973). He has become capable of moving vast quantities of rock, gravel, and soil from one locality to another and is even able to create new reliefs. He can raise water above its natural channels. Such achievements cannot be brought about by natural processes.

Man's deceleration and acceleration of natural processes

Not all human intervention is directed against natural processes. Some may follow their trend, merely changing their rates; this suffices to disturb the steady-state equilibrium. Sometimes, this intervention can be positive not only from the human point of view but also from the point of view of maintaining the natural equilibrium, e.g., the terracing of steep slopes in areas with critical values of the equilibrium: the Andes, Nepal, and the Mediterranean area furnish examples. In these regions, natural catastrophes, such as earthquakes, exceptional monsoon rain, fire, and others, can disturb the equilibrium, and Man's intervention can prevent this to some extent. Terracing influences the steepness of the slope and thus reduces the probability of erosion. On the other hand, the natural runoff on a steep hill can be accelerated manifold by ploughing along the line of the main gradient.

Concentration of Man's activity on small but intense spots

Most activities of Man pertain to discrete areas, be it in densely built areas, open pit mines, highways under construction, or rivers being dammed. These activities, when compared with those of natural processes, are, however, highly intensive. Their impact per unit area is many times greater than that of natural processes. Whereas the total erosive capacity of the Colorado River is some 700,000 tons of silt per day (Golomb and Eder, 1964), this activity is spread over the entire watershed of the river. Its impact per unit area (if considered to be spread out evenly) is therefore rather small, scarcely eroding a layer a few microns in thickness. This should be compared with the result of human intervention in digging, say, a lock, or building a dam, in such a river.

Destruction of ecosystems

The Earth's surface is made up of a number of interdependent systems, or ecosystems, limited areas where the predominant processes create a particular landscape (forest, desert, lake, etc.). Man's activity, even if not willfully intended, does destroy ecosystems, because of the close relationship between the different elements. The tapping of a spring in a desert area and the conveyance of its waters by a pipeline can spell the end of the existing ecosystem by killing the vegetation, destroying the soil texture, and leaving the area prone to wind erosion which may sweep away its last traces.

1

Man and Forest

THE CLEARING OF FORESTS

The Purposes and Effects of Forest Clearing

"No great civilisation likes forests."
K. F. O'Connor*

The areas of dense population coincide with those zones eminently suitable for forests, such as we know in China and India, with their high population density, and in the highly industrialized Northern hemisphere (Eastern U.S.A., southeastern Canada, Europe including the western and Ural parts of the U.S.S.R.). In these areas, however, woods and forests exist today only where Man permitted. The history of population spread is that of forest clearing; witness all the names of localities including words as "wood," or names of trees, in areas devoid of woodland today.

Wood was in the past the most important source of energy and building material in the temperate and cold climatic zones; to some extent it holds that position even today (fuel, building materials, tools, wheels, ships, etc.). The wood, timber, and pulp industries are flourishing today as never before, manufacturing paper, which is an essential component of our culture and could hardly be substituted by other materials. It seems that the commercial exploitation of forests in the future will be at least as important as in the past.

The clearing of forest, of monsoon jungles, or of wooded savannah has always aimed at creating a *new region for the supply of food.* In ancient times, thick forests offered very little to the hunter or the collector, even less to the agriculturist (Stewart, 1956). By clearing through controlled burning, the game was concentrated and more easily hunted. Later, in more advanced stages, grazing areas were expanded and

* Lincoln College, Christchurch, New Zealand.

fields were created. Settlements favored clear areas to warn them of human or natural enemies. Even today, efforts to settle Amazonia have started by wide-scale deforestation as a first step toward making the environment more friendly to man. Clearing forests can have disastrous effects on the hydraulic resistance of the soil, the erosion of which is protected by a dense vegetation (Smith and Wischmeier, 1962). Vegetation influences the infiltration rate of water into the soil; as the plants grow, the coefficient of hydrographic resistance is increased (Mirtskhoulova, 1974), surface runoff is decreased, and soil resistance to erosion is increased. Vegetation as a whole is a factor whose effects are manifold:

Rain is intercepted by the leaves, mainly by the layer of dead leaves and vegetation on the ground. As much as 70% of a rainfall can be intercepted by the different parts of trees, shrubs, mosses, and grasses (Birot, 1966), regulating the water budget of the soil in different periods.

The effect of forest cover on the hydrology of a region is illustrated by the history of a spring, a natural event which resembles an experiment done on purpose: the Wolf Spring in the Commune of Soubey in southern France (Marsh, 1864):

> A few years ago this spring did not exist. At the place where it now rises, a small thread of water was observed after very long rains, but the stream disappeared with the rain. The spot is in the middle of a very steep pasture inclining to the South. 80 years ago, the owner of the land, perceiving that young firs were shooting up in the upper part of it, determined to let them grow, and soon they formed a flourishing grove. As soon as they were well grown, a fine spring appeared in place of the occasional rill, and furnished abundant water in the longest droughts. For forty or fifty years, this spring was considered the best in the Clos du Doubs. A few years since, the grove was felled, and the ground turned again to a pasture. The spring disappeared with the wood, and is now as dry as it was ninety years ago.

Runoff velocity decreases and the cumulating scouring action of the water diminishes. The surface roughness (dead leaves, twigs, etc.) diverges the flow into many small fractions and inhibits the concentration of the runoff into well-defined rills and streams.

The influence of the forest as a control of runoff is greatest on deep and permeable soils, where retention capacities can be increased by forest transpiration and where infiltrated water can be held in storage.

The relation between runoff and forest cover has been studied experimentally in some regions. In Switzerland experiments indicated flood peaks to be one-third to one-half smaller in a completely afforested area than in 30%-forested areas (Teller, 1968). In the Allegheny Plateau (U.S.A.), where 17 hectares of watersheds were reforested, areas covered with five-year-old trees were compared to others, still in the seedling stage, and a flood peak reduction in the former of 52 to 84% as compared with the latter was noticed.

The effect of roots consists of increasing granulation of the soil and hence of an increase in its porosity, allowing increased water-storage capacity. Furthermore, roots constitute a network holding the soil together, thereby adding to the soil resistance against erosion.

The biological activity, which is essential for development of a mature soil profile capable of bearing nutrients, is encouraged by vegetation and especially by a forest cover (Gourou, 1966).

Man is faced not with a problem of clearing forests but rather with the need for means of doing so in a controlled manner, harming as little as possible the natural steady state in the system.

Already in the year 1900, experimental plots served in Bavaria to study the effects of vegetation on soil erosion.* On three plots, each of a different slope, both covered and denuded, water runoff was measured. Results were as follows:

	Grass-covered soil			Denuded soil		
Inclination	10°	20°	30°	10°	20°	30°
Yield (gram water/ m^2/year)	14	42	51	835	2,368	3,104

These experiments show vegetation cover to be more important than the inclination: the ratio between the covered and nude plots at the same inclination remains approximately constant: 1:59 at 10° inclination, 1:56 at 20° and 1:61 at 30°.

Regarding the influence of the density and type of vegetation, Demek (1969) found the following rates of erosion in the Carpathian Mountains (Poland):

	Erosion, mm/year		Ratios	
Segment of the watershed	Upper	Lower	Upper	Lower
Wood	0.00003	0.00003	1	1
Grass/meadow	0.00075	0.032	25	1,000
Tilled field	0.9	0.47	30,000	16,000

The negative correlation between size of the growth and the erosive rates is evident.

Destruction of Forest by Fire

"No doubt fire is the primary agent which allowed Man to modify Nature soundly."
A. I. Woeikof, 1901

The clearing of forests by fire is probably as old a practice as the history of Man; involuntary fires—induced by lightning, volcanic eruptions, etc.—no doubt occurred even earlier. Man's control of fire spans some 300,000 years in North Africa, Southwestern Asia, and Europe, some 50,000 years in South Africa, and some 12,000 years in North America (Vita-Finzi, 1974). Domestication of animals, commencing roughly 10,000 years ago (Stewart, 1956), enhanced this practice to provide grazing grounds.

* Wollny, 1900, "Einfluss der Pflanzendecken auf die Wasserführung der Flüsse," cited in Woeikof (1901).

Hanno, describing a voyage in Africa in 500 B.C., records: "By day we saw nothing but woods, but by night we saw many fires burning" (cited by Stewart, *loc. cit.*).

Fire is used exclusively in the tropics to extend culturable areas (Gourou, 1966). Growing secondary forests partially mask this; still, its accumulated implications are enormous. Satellite photographs show that in West Africa some 80% of the total area used to be set afire every year.* This practice, disastrous as it is for the natural system, does not destroy soil structure completely, leaving as it does the roots and the larger trees unaffected, so that erosive effects are mitigated. As secondary vegetation—intentional or natural—colonizes the terrain rather quickly, erosion is lessened further. In the less humid zones, as on the fringe of the savannah or in the Mediterranean area, where developed evergreen oak and pistacia forests exist in precarious equilibrium with their ecosystems, forest clearing by fire is by far more disastrous, since it leads to soil erosion and to destruction of the steady state of the environmental system.

The effect of destruction of vegetation by fire can induce formation of dunes in the margins of arid belts. In the desert fringe of Australia, lightning and fires set by the aborigines have increased the area covered by sand movement (Conacher, 1971). $Carbon^{14}$ dating demonstrates fires set by the aborigines as long as 10,000 years ago. Extension of sand dunes in such regions is ascribed to these activities, as is the case in the Central Eyre Peninsula (Smith et al., 1975), where the "A" (upper) horizon of the soil profile was deflated after 1916 by the destruction of the natural growth of shrubs.

Clearing of Forest by Cutting

The practice of forest clearing by cutting aims at procuring either areas for settlement, fields, or pasture, or timber for commercial purposes. The first aim causes by far more destruction than the second, in view of its lack of discrimination. Historically speaking, forest clearing per se passed its climax in the developed world, whereas commercial winning of timber is still developing.

It is generally accepted that forest clearing per se is proportional to the extent of the development of civilization. Speaking on the clearing of woodland in Europe, Darby (1956) states that "perhaps the greatest single factor in the evolution of the European landscape has been the clearing of the cover that once clothed almost the entire continent."

The extent of deforestation has been estimated numerically. In Central Europe, north of the Alps, in the year 900 some 80% of the region was still covered by forests, whereas in 1900 this had gone down to 24% (Collier, 1972).

It would be outside the scope of this book to quote all the evidence on this widespread human intervention in the natural forest cover, but some historical cases, elucidating the problem, should be presented.**

* Dr. M. J. Mortimore, personal communication.

** Readers interested in the history of forest destruction will find important contributions in Marsh (1884), Bennett (1939), Curtis (1956), and Darby (1956).

As early as the Classic period, woodcutting had become a widespread habit, in no way taking forest regeneration into consideration. Forest exploitation in Imperial Rome caused such restrictions on the supply of wood that by the 5th century Roman architecture had to adjust itself to the growing scarcity of this construction material. For some authors (e.g., Jacks and Whyte, 1939), the decline of the Roman Empire is nothing but a story of deflation, soil exhaustion, and erosion. The cutting and clearing of wood was very widespread in Europe until the 12th century, when most of the clearing for settlement had been carried out. From the 14th till the beginning of the 17th centuries, reduced clearing was followed by rejuvenation of the forests (Fig. 2). A new impetus to forest cutting arose in the 17th century, when the growing population demanded more timber for industry and manufacturing vessels, vehicles, and machines. This period of extensive cutting in southern France was described by Marsh (1864):

> The provinces of Dauphiny, Avignon and Provence . . . were very well wooded, and famous for the fertility and abundance, not only of the low grounds, but of the hills, (until) the close of the 15th century. . . . (Whereas in the 17th century there occurred) an increase of prosperity and population in Lower Provence, . . . there was an alarming decrease both in the wealth and in the population of Upper Provence and Dauphiny, although, by the clearing of the forests, a great extent of plough land and pasturage had been added. . . . It was found that *the augmented violence of the torrents had swept away, or buried in sand and gravel, more land that had been reclaimed by clearing.*

Erosion induced by such forest clearing has contributed to the depopulation of extensive areas. The area under cultivation in the Lower Alps, to cite an example, continued to diminish in the 19th century from 99,000 hectares in 1842 to 74,000 in 1852 (Marsh, 1864). In ten years, 25,000 hectares had been lost through torrents and abuse by pasture. The scars of such erosion resulted in creating a veritable badlands landscape, deeply eroded in the layer of the blue marls (Fig. 3).

Forest clearing has been carried out most extensively in newly inhabited regions, such as North America, New Zealand, and Siberia. Woeikof (1901) mentions an experienced forester, who had to travel 700 km north of the Tomsk-Irkutsk track to find virgin taiga, not affected by axe or fire. In Wisconsin (U.S.A.), where in 1831 practically all the area was wooded and uninhabited, construction of new towns had by 1950 removed almost all the forest from the built-up areas (Fig. 4). New Zealand could be a case study of the irresponsibility of Man for wasting landscapes by forest clearing. Few countries had such a magnificent heritage of forest wealth and sacrificed it for so little gain (Condliffe, 1936). Before the arrival of the Maoris, a thousand years ago, practically all of New Zealand was forested. By the year 1840, however, when massive European settlement started, 36% of the forested land had already been converted into permanent pasture land. Extension of the inhabited area after 1840, especially at the end of the 19th century, was responsible for the worst destruction of the forest. The area of woodland decreased from 160,580 square km in 1840 to 134,680 square km in 1880, 119,140 square km in 1890, and, finally, to 82,880 square km in 1900. From 1840 to 1900, 50% of the forest was destroyed. Small wonder that this highly deforested area requires massive antierosive means; the situation is difficult especially on the South Island.

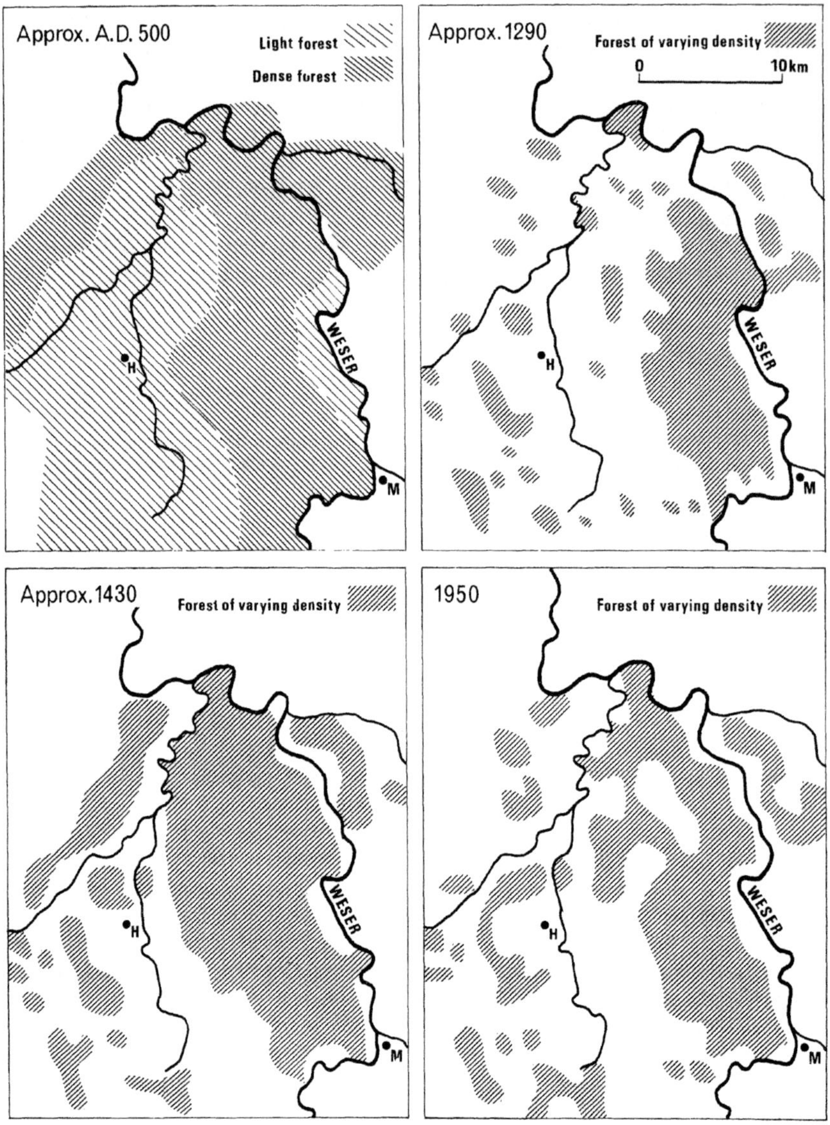

FIG. 2. Changes in woodland in the upper Weser basin, Germany (Darby, 1956)

FIG. 3. Man-made badlands, Mount Saint Michel, Alpes-Maritimes (France), in 1899 (Eaux & Forêts)

Biostasy versus Rhexistasy

Two basic stages can be distinguished in the life of biological systems: biostasy and rhexistasy. Biostasy is the "normal," i.e., steady state of vegetation (and other biological entities) when they are in equilibrium with their environment. Once this equilibrium is disturbed, a new steady state can be achieved by rhexistasy, i.e., by a process of adaptation of the different biological factors to the new situation through elaborate interaction, leading to a new steady state. It is of interest to look into the stage in which the present forest cover is, taking into account Man's intervention as described above.

In 1976 the world's forests occupied 32 % of the land area, or 4,145 million hectares (FAO, 1977). The standing volume of timber is 238 milliard m^3, from which 2.5 milliard m^3, or 1.0%, are cut annually.* This amounts to a worldwide annual yield of 0.66 m^3 of timber per capita. Average values, however, can be misleading; whereas in South America only 0.25 % of the timber volume is exploited yearly, the corresponding values are 0.5 % in the U.S.S.R., 1 % in North America, 2.6 % in Asia, 3 % in Europe, 5 % in Africa, 5.2 % in Central America, and 7 % in Oceania. Most countries

* In 1975, 2,448 milliard m^3, in 1976, 2,498 milliard m^3 (*United Nations Statistical Yearbook*, 1977).

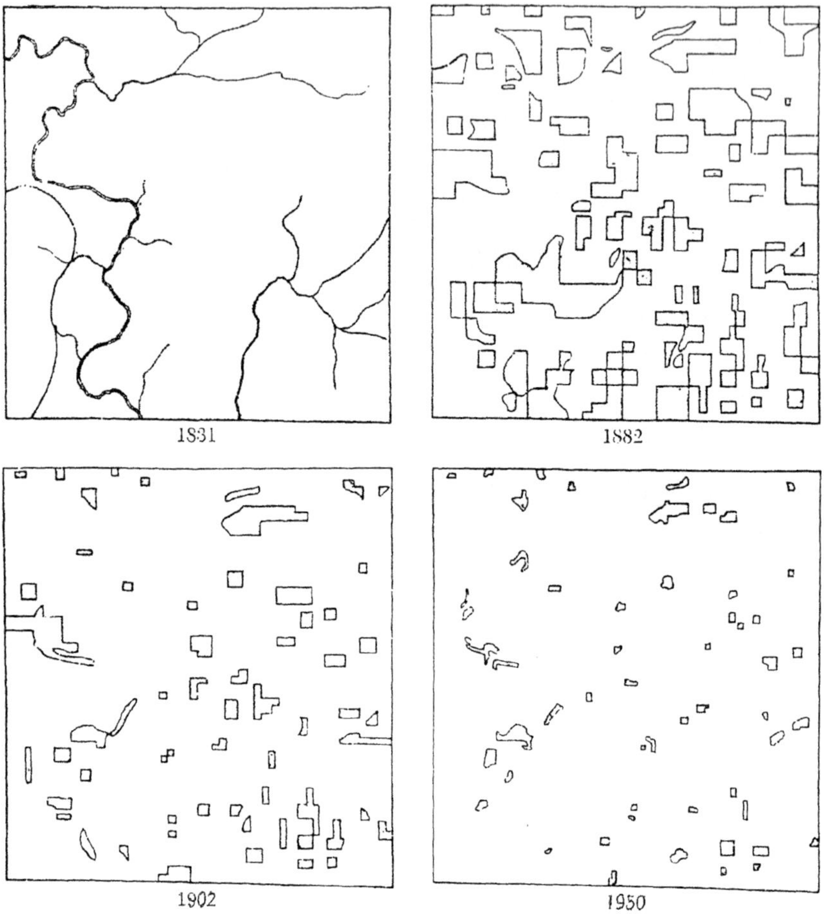

FIG. 4. Clearing of forest in Green County, Wisconsin, 1831–1950, for purpose of urban settlement (Cadiz Township) (Curtis, 1956). The area in the figure measures 10 × 10 km; it is drained by the Petaconica River

are exploiting the forest far above the rate at which natural growth replaces the crops harvested. It seems that the tropical forest is the most endangered: 13 million hectares are cleared annually (*To the Point*, 22.3.1976), amounting to 0.3% of the global forest area.

The perception of the need for forest management and reforestation is growing; evidence from different countries with histories of forest destruction is reassuring. The area of forests in France, estimated at 17 million hectares in 1750 (Marsh, 1864) and a mere 9.3 million hectares in 1860, attained 11.4 million hectares in 1953.* In Japan, as early as 1897 a Forest Act was put into effect (Lowdermilk, 1927). In New Zealand, a countrywide effort is being made by reforestation with the *Pinus radiata* tree. In the United States, where annually millions of hectares are accidentally burnt (in 1972 alone, 810,000 hectares were involved; *United Nations Statistical Yearbook*, 1975), new forests are planted (in 1972, 650,000 hectares). The wide spectrum of states which are reforesting their delicate and erosion-prone regions (Spain, Cyprus, Israel, Wales, Australia, Argentina, Peru (FAO, 1972)) indicates a new, rhexistatic stage of the forest environment.

CHANGE OF SOIL PROFILE BY FOREST CLEARING

Effects on Permeability and Density

From a pedological standpoint, deforestation in temperate zones concurrent with mixing the upper horizons results in changing the original podzolic soils into acid brown forest soils (Yaalon and Yaron, 1966). The deforestation in tropical humid regions results not only in high erosion and dehydration of iron oxides but in a conversion of the original latosoils into ferrallitic crusts (or "laterite crusts"), if the nude state extends long enough and no rhexistatic rejuvenation of the vegetation occurs.

The efficacy of forest cover in absorbing direct rainfall, and the increase of runoff after its clearing (discussed above pp. 16, 17), have also been experimentally demonstrated by Hornbeck et al. (1970). Two parallel watersheds in central New England, of similar slopes of 20–30°, one of which had been cleared of its forest cover, were compared. The soil consisted of a permeable sand and sandy loam, to an average depth of 1.5 meters. Observations over a two-year period revealed that the cleared watershed yielded 34% more runoff than the covered one. As the forest floor on the cleared plot remained intact, no erosive effects occurred. It seems that not the act of cutting itself but the removal of trees, shrubs, and humus changes the equilibrium of the forest system; after all, fallen trees are present even in the virgin forest. It is the practice of commercial deforestation, employing heavy tractors and logging trucks, that is responsible for changing the porosity of the soil by compressing it and by making it impermeable, which is the first step to overland flow and erosion.

An investigation of the effects of tractor logging on the physical properties of soils has been made in southwest Washington (Steinbrenner and Gessel, 1955). Nine areas were selected for measuring the effects of cutting practices. Values were obtained from the undisturbed woodland (as a control), from tractor roads, and from tractor cutovers on each of these areas. The soil was examined to a depth of 7.5 cm. The results are rather convincing: permeability of the soil (in this top layer) decreased by 35% in the tractor cutover (as compared with the undisturbed area) and by 93% in the tractor road, making the roads practically impermeable. The bulk

* D. Lowenthal, editorial note to the 1965 edition of Marsh (1864).

density increased by 2.4% in the tractor cutover and by 15% in the tractor road; macroscopic pores decreased by 10% in volume in the cutover and by 53% in the tractor roads area. As the roads take up 26% of the working area, a quarter of the exploited forest is converted to a dense, impermeable region where water and melting snow cause runoff and erosion.

Effects of Vegetation Clearing on The Soil Salinity

Man-induced soil salinity has mainly been considered as irrigation-caused (pp. 49, 50). Soil salinity in nonirrigated agricultural areas, a documented phenomenon, has received relatively little attention (Conacher and Murray, 1973). It is a fact that clearing of forests, where a precarious equilibrium between the vegetation and its environment exists, has—under adverse climatic conditions—caused salinity even where no irrigation took place. Western Australia's wheat belt, a forest two generations ago, is one of these critical areas.

Salinity appears as *salt scalds* on bare soil containing high concentrations of soluble salts, mainly sodium chloride (Conacher, 1974). The phenomenon has been reported from other parts of Australia (South Australia, N.S. Wales, Victoria) as well, but is most severe in Western Australia, where its total extent was estimated in 1970 at 162,000 hectares, an increase of 122% over 15 years.

These salt scalds formed, according to Conacher and Murray (1973), after European settlement had begun, and are quite distinct from salt lakes or playas, which are of natural origin.

Conacher and Murray's study was carried out on 36 farms in an area of 1,000 square km east of York, on the Avon River, some 120 km east of Perth. Only three of the inspected farms had no salt scalds; in the others, an average of 2.83% of the farm area was affected, the average scald area per farm being 34.4 hectares. Most scalds were associated with erosion caused by rills of less than 30 cm deep and were situated in the valleys or on their slopes (Fig. 5). All these farms had over 70% of their area cleared, with less than 15% still under natural vegetation.

The clearing of vegetation in the area was largely completed by the 1930s. There is, however, evidence of scald expansion since 1962, and new scalds appeared as recently as 1965. We may therefore assume that neither a rising water table nor increased pressures in confined aquifers had caused the scalding, since these latter factors have not changed over a long period.

Salt scalds formation is connected with the particular runoff, which increases following the clearing of vegetation (pp. 30–32; see also Kedar, 1957a, b). Western Australian soils contain soluble salts (from near-zero to 450 ppm); runoff removes these salts and redeposits them in low-lying areas which become salt scalds, where the NaCl content may reach 3,000 to 46,000 ppm. A static situation in dry years will change to a fluid one in wetter periods. Salt accumulation is assisted by a low surface gradient (less than one degree in most of the cases). It seems that the combined effect of runoff, erosion, and apparently seepage is the most important factor in the development of the scalds. The supposed development of the scalds is represented in Figure 6. It begins with a presumed low salt content prior to the clearing, followed by

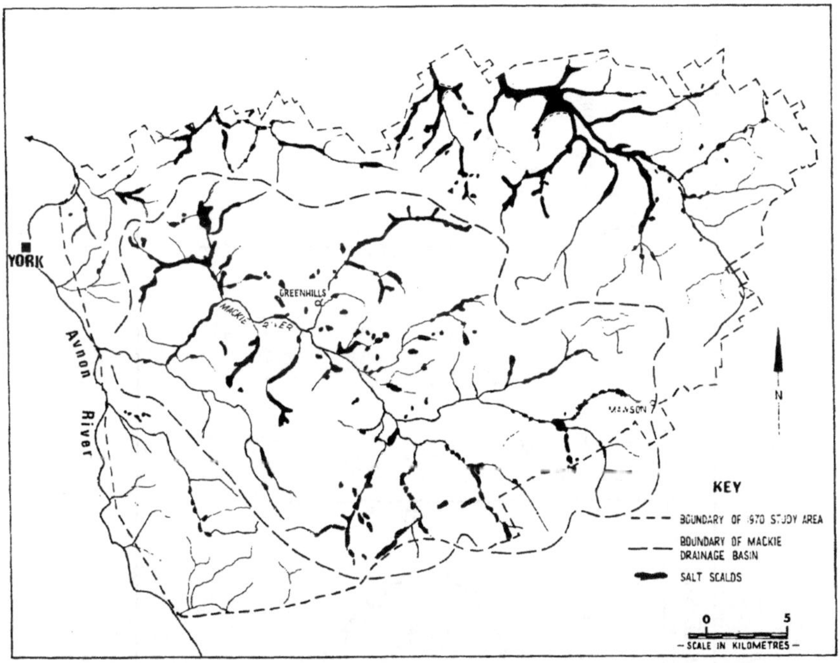

FIG. 5. Salt scalds in the York-Mawson area, Western Australia (Conacher and Murray, 1973)

increase of salinity with the removal of natural vegetation. Two trends are now possible in further development: either attainment of a new steady state with a relatively high salinity or a return to the initial steady state after all the salts from the area have been drained to the sea. The evidence of increase in the number and surface of scalds suggests that the first trend prevails.

A new conservation intervention of Man could, however, lead to another possibility (Conacher, 1974). Construction of intercepting banks on the slope leading to the valley bottom can redistribute the salts to form concentrations tolerable for a vegetation-carrying soil.

Effects of Destruction of Tropical Forest: Accelerate Laterization

Laterization is a process, by weathering, of accumulation of iron and alumina, together with the leaching of silica and bases (Mohr and Van Baren, 1954), in a wet, hot tropical environment. Sometimes the lateritic soil occurs in the form of solid rock or a hard crust and has been explained as a fossil horizon C of ancient, eroded soils. On the other hand, there are evidences of active formation of this type of hard crusts ("carapace," "cuirasse"). It has been proposed that at least part of the actual

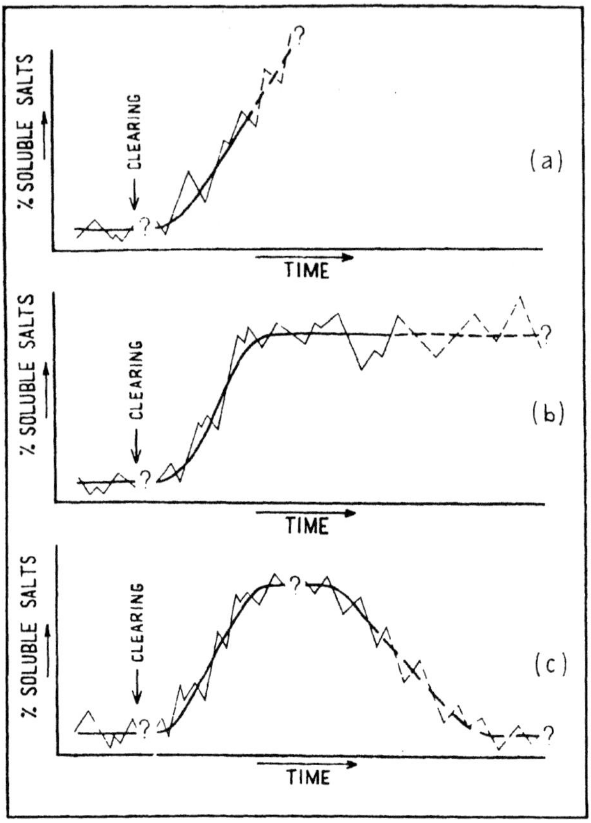

FIG. 6. Three possible trends in the development of the salt scalds in Western Australia. Broken lines suggest future evolution (Conacher and Murray, 1973)

formation of lateritic crusts has at its origin an anthropic stimulant (Duchaufour, 1965).

The red lateritic soils—perfectly developed in three horizons—exist under the cover of equatorial rain forest; they are in a steady equilibrium with the climate and the vegetation. The soils on the margin of the equatorial rain forest, however, with a marked dry season, are in a more delicate equilibrium with their environment, and cannot withstand the destructive action of man. This destructive action is manifested either by an exaggerated shifting cultivation or by the use of fire, or by both.

The cultivation of a deforested field parcel is limited to two to four years, and after that the field is abandoned for two to four decades, and the soil renews its original structure in this period by the spontaneous vegetation which captures the abandoned field. But if the period of rest is less than 10 to 20 years, the structure of the soil is being destroyed, the humus is being leached, and the microclimate of the soils changes in the direction of progressive dessication in the dry season. As the beginning of a cultivation is generally accompanied by a deliberate bush fire, the soil becomes even harder.

As the forest is progressively replaced by savannah, the degradation continues: in the dry season, the superficial horizons are evacuated by wind, and horizon B is exposed to erosional processes. In a horizontal environment, the water table nears the surface, and a concentration by capillarity of ferrous minerals becomes evident; horizon B is being invaded by iron oxydes. When these dry up, they become hard, lateritic crusts, which are entirely sterile and are neither of agricultural use nor suitable to produce any natural vegetation.

EROSION IN DEFORESTED AREAS

Topographical and Climatic Controls

The strongest erosive effects occur when forests are stripped in areas of great relief energy (steep slopes) or of great potential runoff (high intensity of precipitation). Highly mountainous areas, semiarid or equatorial, prone to monsoon rains, are especially sensitive. Deforested lowlands in temperate zones show milder erosive effects.

The influence of the removal of forest cover is therefore the greatest in the catchments of the lower orders, i.e., in the highest reaches of a watershed. In the U.S., about 56% of the current average annual floodwater and sediment damage occur in valleys of low-order streams. Forest canopy destruction on steep slopes results in increased peak flows, as demonstrated by analyzing data from 54 watersheds in the Cascade Mountains (Teller, 1968): clearing 1 square km increased flood peaks by 0.5 $m^3/sec/km^2$.

Deforestation of tropical lands, and its destructive effects on the environment, although treated by Gourou (1956, 1966), still suffers from a lack of numerical evaluation. One study on the Cinamuk River (Java) showed the direct effect of deforestation: at an average annual rainfall of 740 mm and a maximum natural discharge of 1,300 m^3/sec, clearing of some 17,000 hectares of the forest (equal to 5% only of the watershed area) increased the runoff between 30% to 70% (Partosedono, 1974).

Another study, in semiarid southern California, compared the Monroe Canyon, where 17 hectares woodland were cleared (in the years 1958–1959), with the adjacent Wolf Canyon, where the wood was left in its natural state. After a fire in 1960 striking both watersheds, Wolf Canyon was allowed to revert to its natural state, whereas around Monroe Canyon a further 57 hectares were cleared. Measurements between 1963 and 1969 showed that the volume of sediment per area unit carried by the Monroe Canyon was eight times greater than the volume carried by the Wolf Canyon (Orme and Bailey, 1970).

Effects of Deforestation on the Erosion of Slopes

Evidence of a "cause and effect" relationship between deforestation and mass movements on slopes, such as landslides, earth flow, slips, etc., seems to have been known long ago. Marsh (1864) mentions many examples in southern France, which occurred between the years 1707 and 1768, and which — in his opinion — clearly prove the case.

The stability of a slope depends on a balance between forces: the slope begins to move when the shear forces on any potential failure surface exceed the shear strength. It is expressed as a factor F:

$$F = \frac{\text{shear forces (promoting sliding along a critical surface)}}{\text{shear strength (along that surface)}}$$

(Gray, 1969). Soil constitution, slope granularity, and pore water pressure are the primary factors that determine the behavior and the stability of slopes.

When the soil on a slope becomes saturated by water, friction and cohesive forces are reduced to the point that they cannot withstand gravitational forces, and the value of F becomes critical. In the soil of a covered slope, free water contents are reduced through the action of the roots, and cohesion accordingly increases.* This improvement is, of course, not unlimited.

A clearing in the Cascade Range (Oregon, U.S.A.) in 1961 led Gray (1969) to classify 47 events of earth movements, the debris of which amounted to some 243,000 m^3. Results were as follows:

Classification A: By character	*No. of events*
Earth flow	18
Slump	10
Earth flow through channel scour	5
Slump through channel scour	4
Debris slide through channel scour	2
Channel scour	3
Roadfill washout	3
Avalanches	2
Total	47
Classification B: By cause	*No. of events*
Roadfill failure	12
Roadfill backslope failure	5
Roadfill and backslope failure	6
Road drainage water	8
Roads removed by streams	3
Events in logged areas	8
Events in undisturbed areas	5
Total	47

* Some of the best timber often occurs on high unstable slopes (Bailey, 1973), as the presence of loose deep soil with high moisture is conducive to forest growth.

Analysis of other data shows that only 6.4 % (by volume) were dislodged in the logged area, whereas in the undisturbed area 40.6 % (by volume) were dislodged. The bulk of the disturbance was caused by the accessory means of production—the roads, and especially the unadequate drainage of the roads (25 % of all volume involved).

The role played by accessory roads in the intensity of soil erosion of logged forest is a most decisive one: the 5 to 15 % of the logging area usually covered by roads produce some 90 % of the sediment. Experiments in the Fernow Experimental Forest (West Virginia, U.S.A.) showed that turbidity from a watershed from which the trees had been cut, having neither planned roads nor provision for their drainage, reached 56,000 ppm; with unplanned roads, but drained, however, this amounted to 5,200 ppm; with moderate planning and drainage, 210 ppm; with careful planning and drainage, 22 ppm. For the untouched area, the natural turbidity content was 15 ppm (Lull, 1971).

The direct effect of woodcutting on the slope is the decrease of the stress produced by the living tree, which—for Douglas firs—has been evaluated at 500–750 kg/m^2, with a total base area of 75 to 115 m^2 per hectare (Gray, 1969). On cutting, the decrease of timber weight reduces the shear strength of the soil under the tree root system (Fig. 7), resulting in decompaction of the soil. Loss of the root system, which constitutes a strength-building factor in the soil, contributes to accelerated erosion after the logging.

FIG. 7. The anchoring effects of the roots. When felled, the mass included between the roots is prone to erosion. Bagley Wood, Oxford

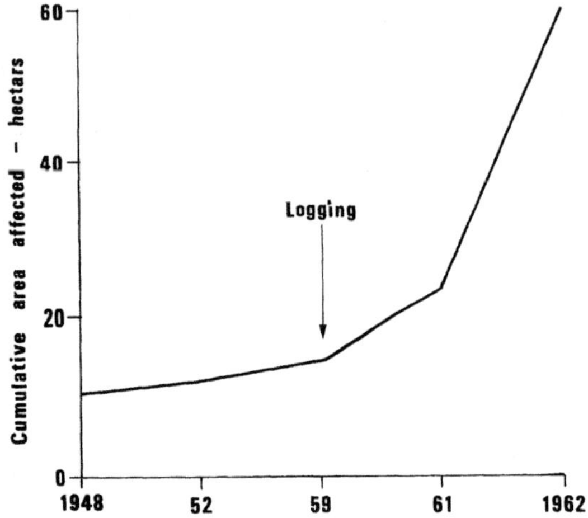

FIG. 8. Acreage of slides before and after logging. Maybeso Creek, Alaska (Bishop and Stevens, 1964)

Additional evidence of landslides induced by the logging and clearing of forests has been brought from Alaska (the Maybeso Creek; Bishop and Stevens, 1964). Aerial photos prior to 1948 showed 12 old scars of landslides, of an overall area of about 11 hectares. Between 1948 and July 1952, one more landslide occurred, covering 1.7 hectares. These 13 old scars, triggered by natural causes, represent a period of steady state practically unaffected by man.

In 1950, lumbering, formerly limited to the valley, began reaching the slopes. New landslides, as shown on aerial photos, were as follows:

July 1950–May 1959; 20 events, 2.5 hectares

May 1959–May 1961: 28 events, 8.6 hectares

May 1961–July 1962: 68 events, 36.3 hectares

The rapid increase after 1950 is obvious (Fig. 8); the area involved has increased some 4.7 times.

The destruction of not only forest but even of grass exposes soil to erosion, as illustrated in New Zealand (Eyles, 1971). Investigations at the Tangoio Experimental Farm near Napier showed that, out of 52 sample profiles in an area totaling 162 hectares, 40 were related to movements — slumps, slips, and slides. This slope erosion connected wit the change-over from forest to shrub, was enhanced by removing shrub to make place for grass for grazing purposes.

DEFORESTATION AND SEDIMENTATION

Erosion and sedimentation as a result of deforestation are the most ubiquitous and notable of all the processes resulting from deforestation (Bennett, 1939). Soil con-

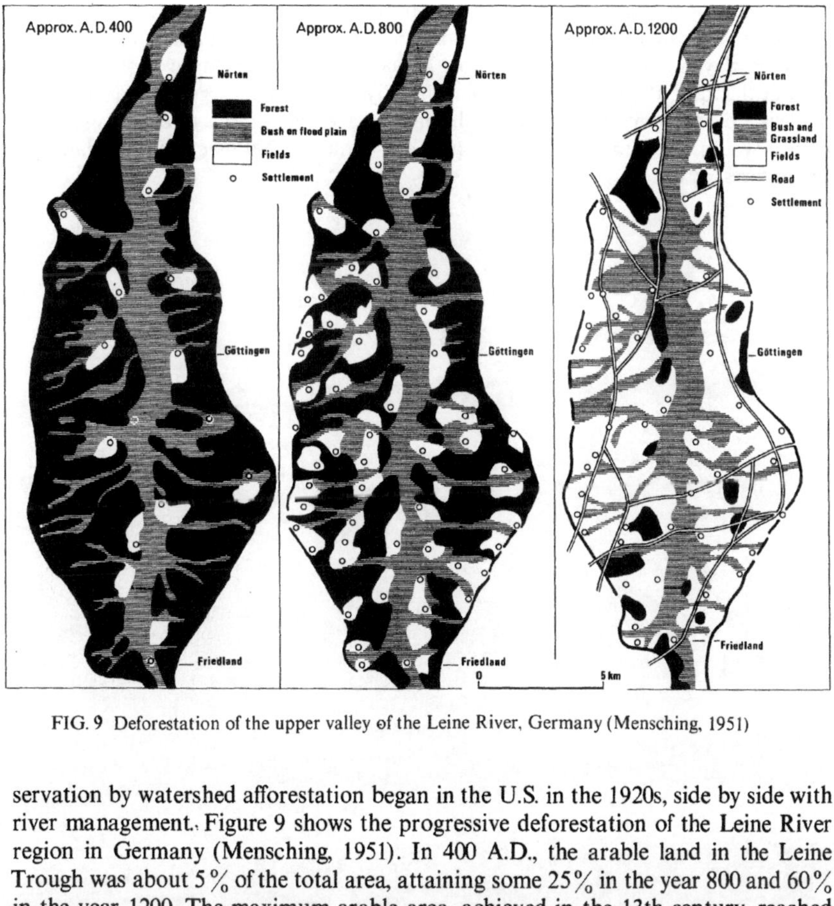

FIG. 9 Deforestation of the upper valley of the Leine River, Germany (Mensching, 1951)

servation by watershed afforestation began in the U.S. in the 1920s, side by side with river management. Figure 9 shows the progressive deforestation of the Leine River region in Germany (Mensching, 1951). In 400 A.D., the arable land in the Leine Trough was about 5 % of the total area, attaining some 25 % in the year 800 and 60 % in the year 1200. The maximum arable area, achieved in the 13th century, reached 75 %, subsequently decreasing to present-day values of 40 %. During the deforestation period, the young and fine clays and silts ("Auelehm") accumulated. Forest destruction is blamed to have triggered soil erosion from the mountains, with subsequent deposition in the coastal plains or on the Mediterranean shelf. Vita-Finzi (1964) considers that most of the lower terraces of the rivers flowing into the Mediterranean were formed in this way.

In wet tropical regions, where dense forests may cover up to 30 meters of soil, deforestation causes heavy erosion and formation elsewhere of sediments (Douglas, 1967a). Studies in tropical Australia (North Queensland) showed a catchment with 99 % of its area still under natural vegetation to have a sediment load of 21.1 m^3/km^2/year, whereas neighboring ones, where partial deforestation left only 64 % of the surface covered, yielded 103.1 m^3/km^2/year, i.e., five times more.

The rates of erosion induced by forest destruction can be illustrated well from a comparison by Bennett (1939). Under a forest cover, sediment loss on a 10° slope of sandy clay is 0.005 ton/hectare/year. In, say, 500,000 years, a layer of 17.5 cm is lost; at this rate, the loss can be replenished by the decay of organic and inorganic material. Under similar climatic conditions, an area growing a tilled crop such as cotton would lose those 17.5 cm in 46 years, a rate *10,900 times greater* than that under forest cover.

2

Pasture and Grazing

INTRODUCTION

Grazing by domesticated animals is not only one of the most ancient but also one of the most persistent occupations; it is as important today as ever, and it is is by no means on the decrease. Only its methods and scope, of course, have changed with modern commercial land use. Just as with other rural land uses, the diversity of methods employed is a very great one, from the nomadic ones of the Sahel to modern ranching in the Southwest U.S. and in New Zealand.

In 1976, approximately 1,036 million sheep and 1,212 million cattle occupied 3,058 million hectares of pastures and meadows (FAO, 1977; *United Nations Statistical Yearbook*, 1977). These numbers, approximate as they are, describe the scope of the problem. Remembering that the creation of grazing land by destruction of forests or by burning shrubs or savannah was ever widely practiced, its impact on the development of soil and slopes will be evident.

Land use through grazing can take place in harmony with the natural steady state if the equilibrium between growth and consumption and the balance of the nutritional and textural state of the soil are not disturbed. This concept of equilibrium is expressed pragmatically by the *land-carrying capacity* (Stocking, 1972): any area of land will support over long periods only a limited number of stock. An *absolute* limit is imposed by soil and climatic factors, and a *practical* limit is set by the way in which the land is used.

The problem is, therefore, not the grazing by itself, which, of course, introduces anthropogeomorphological agents into the terrain, but that of *overgrazing*, which is of a more economic and sociological nature. Overgrazing includes overpopulation of animals per unit area, compaction of the soil on gathering places, such as points of water supply, and the introduction of grazing to terrains the vegetation climax of which is very precarious, such as in deforested areas on the semiarid fringe of humid pasture lands.

PROBLEMS OF OVERGRAZING

The Ratio Livestock/Grazing Area

Of all factors, the ratio livestock/grazing area has the greatest impact on the use or misuse of the pasture land. This ratio varies considerably, being 3–4 heads/hectare in the Sahel (Dresch, 1977), 2.5/hectare on the ranches of the Southwest U.S. (Hadley, 1977) and 0.25/hectare in Australia (Perry, 1977). As in developing countries grazing areas generally belong commonly to a village or a tribe, there is no individual incentive to keep them in good condition; on the contrary, they are exploited as much as possible (Pereira, 1977), as the population will depend, more or less completely, on this form of livelihood. Without enforceable legislation to guard the common grazing land properly against erosion, no solution to overgrazing and to accelerated erosion seems to exist.

Overgrazing sometimes means also introducing a herbivorous animal, which destroys the grass, and hence the natural equilibrium, by its great consumption. One case is the "man-made desert" in Central Otago, New Zealand. The region receives annually 500–600 mm of rain, falling mostly in the warm season, and is therefore at the fringe of the grassland ecosystem. A century ago there flourished a tussock-grass vegetation of moderate density. Today the situation is worse: "The European man allowed sheep, *rabbit* and fire to create their combined devastation, which resulted in creation of one of the world's most desolate deserts" (Clark, 1949).*

The Paths Followed by the Gathering of Animals

The paths leading to the watering points are the initiating lines of erosion. Hard in the dry season, muddy in the wet ones, they constitute the grassless segments of a pasture land, changing—in accentuated terrains—to gullies which in time become unsuitable for tracks. Abandonment is followed by formation of a new strip, parallel to the former, of a similar nature. In the Sahel, as many as 6,000 cattle or 50,000 sheep and goats may gather daily near a water hole (Nir, 1974). In Rhodesia, all cattle have to be dipped in a communal tank once a week, to control tick-borne diseases, each tank serving some 17,000 cattle (Stocking, 1972). As the tanks are dispersed, 1 tank on 100–600 km^2, trekking distances of 20 km may be involved. The cumulative effect of cattle gathering and dispersing triggers gully formation and wind erosion (Fig. 10). Another example of erosive processes induced by concentration of sheep in a limited area is from the Welsh uplands (T. M. Thomas, 1956). The rubbing effect of sheep against the steep slopes has led to the removal of 122,000 m^3 of soil from an area of 18.2 hectares in a sample area of 364 hectares, as they created *steps* for themselves. This is, if extrapolated to a general scale, a removal of 20,000 m^3/km^2 from the eroded area, and over 1,000 m^3/km^2 from the sample area.

* This statement describing the area as being one of the world's most desolate deserts seems somewhat exaggerated.

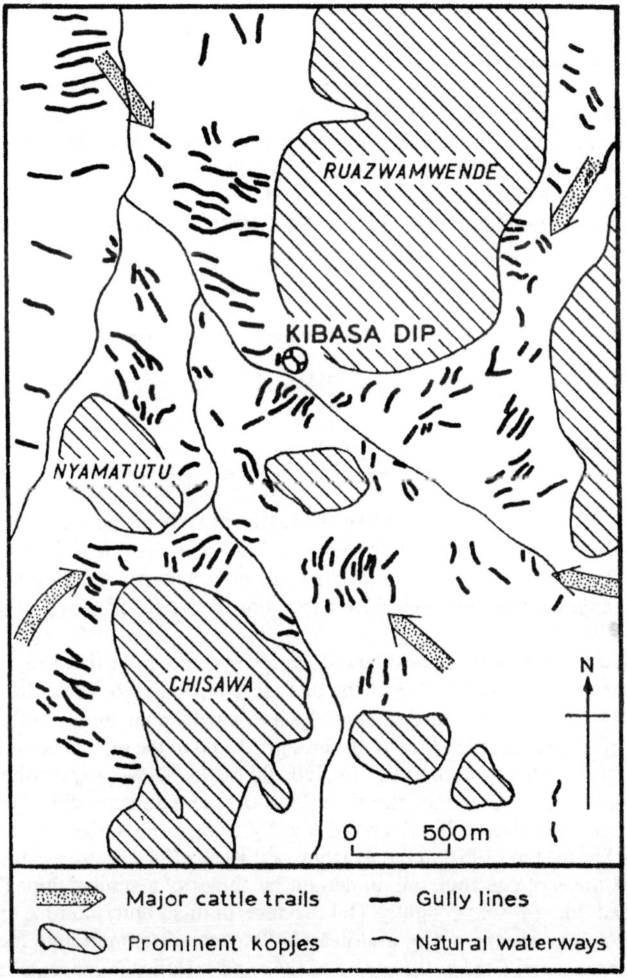

FIG. 10. Gully erosion induced by cattle, Zimbabwe (Stocking, 1972)

GEOMORPHOLOGICAL EFFECTS OF OVERGRAZING

One of the results of overgrazing is the change of the soil profile (Yaalon and Yaron, 1966). The "A" horizon degrades, although the mineral contents are enriched by the manure. The change depends on the slope of the original soil, the climate, and other factors. In the Mediterranean and in semiarid environments, grazing can change the original lithosoil into a calcareous crust.

Overgrazing has often been considered the main factor in the sedimentation occurring in the lower reaches of rivers, especially in the semiarid zones of South Africa and the Southwest U.S. Talbot (1961) states that the present erosive state of the Oranje-Vaal basins was formed during the period 1880–1930 as the grassveld was both partially burned away and overstocked by cattle and sheep.

An important study on the effects of overgrazing was made in the Colorado Basin (Hadley, 1974). The amounts discharged and settling were measured in four stations. Two periods were compared: 1926–1941 and 1942–1960. The ratios between the data of the two periods, (1926–1941)/(1942–1960), at the four stations were as follows:

Station	Ratio of average discharges	Ratio of average sediments
A	0.944	0.486
B	0.973	0.501
C	1.156	0.683
D	0.775	0.381

Examination of the ratios shows that the sediment yield in "period two" was much lower, about one-half, of that of "period one," whereas the range in the discharge did not differ by more than 0.225 from the value of 1. Hadley attributes this phenomenon to changes in the land use, since variations in the runoff played a minor role only.

This watershed is mostly arid grassland (87%) and little forested (13%). The forested area contributed 94.4% of the discharge but only 16.7% of the suspended sediment. It is therefore the arid grassland which chiefly contributes to the sediment load. The main change in land use of the arid grassland in the second period consisted of a reduction of livestock number by 750,000 cattle. Grazing control can, as the case study points out, reduce the runoff by 25% and the sediment yield by nearly 50%.

The difference between "light" and "heavy" grazing and its effect on erosion was studied by Sharp et al. (1964) in the Cottonwood Range Field Station, South Dakota. Here, three areas of chestnut soil, underlain by shales of a similar slope (7.6–7,9%) were grazed for 14 years, lightly (1.4 hectare/animal unit/month), moderately (1 hectare/animal unit/month), and heavily (0.6 hectare/animal unit/month). The experiment was conducted in the spring of 1963 and included four events of rain. The runoff from the various areas was as follows: lightly grazed, 17%; medium grazed, 20.9%; heavily grazed, 23.8%. That runoff increases with increasing intensity of grazing is thus shown clearly once more, leading ultimately to increased erosion.

In light of these facts, which attitude should be taken towards grazing economy as a "genre de vie"? Glenn (1911) stated

> when properly cared for and not too steep, grazing land should be capable of indefinite use. The erosion of grazed lands is due chiefly to ignorance of the impact of the overgrazing and neglect. The remedy is not reforestation but education.

3

Geomorphology of Agriculture

METAPEDOLOGY

Of the Earth's land surface, 11.4% — 1,488 million hectares — are tilled agriculturally (United Nations Food and Agricultural Organization, 1974) and have been cultivated, ploughed, and sown for periods ranging from some tens to thousands of years. In spite of considerable industrial development, agriculture's role is increasing, supplying food and raw materials for industry. The fundamental properties of tilled soil, and its structure and mineral content, change qualitatively and quantitatively at a much higher rate than would happen by natural processes only. This man-induced change in soil properties has been coined "metapedology"(Yaalon and Yaron, 1966).

Soil can be agriculturally characterized as an open system, which includes biological, topographical, lithological, as well as human factors (ploughing, irrigating, draining). Man's action may have beneficial or detrimental effects; modern technology can increase soil fertility remarkably, but it also can accelerate soil erosion.

The influences of Man on the lithological content of the soil were known from ancient times: lime, to improve meager soils, was used in France in Roman times (Bidwell and Hole, 1965). Peruvian guano was used in England in the 1820s, and Chilean nitrate in Virginia in the 1840s. Removing of salts by deep drainage has been practiced in some semiarid regions (Nir, 1974), whereas soil salination by excessive irrigation was known in Mesopotamia in the second millenium B.C. In the Highlands of Britain there are soils made through addition of turf, peat, and/or seaweed to sandy, unfertile soils (Bidwell and Hole, 1965). The addition of sea sand to peat was practiced in Ireland already before the 4th century A.D.

The capabilities of Man to improve soil characteristics are rather limited. By his intervention he may hasten or slow down natural processes (leaching, acidification, podsolisation, gley and peat formation) but he cannot counteract the transformation from one pedogenetic system to another.

TECHNIQUES OF SOIL CULTIVATION

Ploughing

The main techniques of soil cultivation have remained unchanged since times immemorial, differing only in the size and rapidity of the incisions and the overturning of the soil. There are two main methods of soil cultivation: one, the spade, the hoe, or one of their variants, which overturns the topsoil in isolated scars, and two, the plough, which does the overturning in continuous furrows. The methods represent no hazard per se; choosing one or the other depends on the technology available. The hoe is still the main tool in the wet tropics, not only because of tradition and cheap labor but because it is better adapted to the environmental conditions imposed by heavy rains, especially when cultivating crops other than cereals. Replacing tools suitable for a particular environment by others for apparent economic or technological advantage may result in destruction of the equilibrium achieved previously. Bennett (1939) blames the new, European agricultural practices, introduced in different parts of the Americas, as having caused accelerated erosion. Maize, the main native grain crop, was planted by the Indians on rather steep slopes, not in continuous furrows but in groups, and only the spade or hoe were used to aereate the soil. The same holds for tobacco and potato. The plough, introduced by European farmers of British lowlands' or Spanish huertas' background, promoted erosive processes. Recent cultivation in so-called "new" countries by unchecked ploughing has caused similar effects.

The effects of ploughing depend heavily on the angle between the direction of the furrow and that of the slope. Ploughing from the top of a slope vertically to its foot may technologically represent the simplest manner; it certainly is widely used. This mode is, however, the most dangerous manner of soil cultivation; the furrows become rill, gullies, and carriers of runoff. *This type of intervention is perhaps the single MOST IMPORTANT AGENT OF EROSION induced by man.*

Another method of ploughing which takes the slope into account is contour ploughing. Compared with vertical ploughing, this method is most effective on medium slopes of deep permeable soils; its effectiveness decreases as the slope becomes either very flat or very steep (Smith and Wischmeier, 1962). The ratio of soil loss by contour ploughing and that from vertical ploughing increases with the slope angle: up to 7‰, the ratio is 0.5; for 7–12‰, 0.6; for 12–18‰, 0.8; and for steep slopes of 18–24‰, it increases to 0.9. In other words, on such steep slopes the soil loss is nearly equal for both methods, but still contour ploughing is preferable. No wonder that contour ploughing has become one of the most important tools of soil conservation.

Another problem of ploughing concerns the influence of its depth on the erosion. Ramser (1917) holds greater depth to be beneficial, as increased porosity (from deeper furrows) will increase water storage capacity, thereby lessening erosion. For example, 25 cm of loose soil will absorb 50 mm of rainfall.

The problem of deep ploughing is, however, more complex, depending, inter alia, on the time elapsed between ploughing and becoming covered by crops. The shallow Mediterranean furrow has withstood erosion much better than the deep furrow imparted by modern ploughs, which is better adapted to the deep soils of the steppes and prairies. The shallow furrow permitted moisture to remain, keeping it for the growing young plants.

It was in the Midwest of the United States that huge tracts of land were converted from grazing areas to food-growing agriculture. There exists ample evidence that uncontrolled and environmentally harmful methods caused river and slope erosion. McLaughlin (1947) and Schumm (1963) studied the Cimmaron Valley, Kansas, where accelerated erosion followed a disastrous flood in 1914, destroying a bridge, buildings, and fields. At the time of study, the width of the river ranged from 47 to 1,000 meters, whereas in 1831, 1847, and even 1905, witnesses had described the river as a "narrow, running stream," generally no more than 7 meters wide. At one point, where the road crossed the river, its width was given as 20 meters in 1874; in 1916, after the flood, it had attained 215 meters. Several authors (quoted by McLaughlin) showed the greatest incidence of channel erosion in the Midwest to have occurred between 1880 and 1900. Before 1914, not much bed material load could have moved through the narrow, meandering channel (Schumm, 1963). The strong erosion after 1914 should be explained by a local factor. This factor seems to be the rate of ploughing in the area (McLaughlin, 1947): in Kansas, the average annual rate of newly ploughed area, which was 62,000 hectares from 1888 to 1903, rose rapidly and by 1914 reached 312,000 hectares. The environmental factors came on top of the new situation, where grassland was carelessly replaced by cultivation, with soil erosion as the inevitable result. Measurement of channel widths by aerial photographs in 1960 (Schumm and Lichty, 1961) suggested only a slight widening in the period from 1954 to 1960. The stabilization of the flood plain is credited to the changing vegetation (cottonwood, willow instead of grass) with deeper roots.

One of the most vivid examples of soil erosion caused by changing a natural environment by means of heavy machinery was the groundnut scheme in Eastern Africa in the late 1940s (Wood, 1950; Gourou, 1955). The scheme managers were determined from the start to prevent erosion. Still, the hundreds of tractors employed in the clearing were "raising clouds of dust before and behind them." Moreover, the cleared soil, consisting of 80% gritting quartz sand, dried up after the rainy season and was baked by the sun into "something that was similar to a sheet of concrete." The heavy tractors, passing over the soil, converted it unintentionally into a surface of macadam, from which runoff caused heavy erosion.

It is not the aim of these examples to suggest that modern agricultural technology should be rejected from tropical, Mediterranean, or other critical regions. Rather they illustrate the importance of proper technology as an element in the preservation of the steady state between Man and nature. The importance of technology is growing steadily: witness the number of tractors, over 18 million in 1976 (*United Nations Statistical Yearbook*, 1977). These tractors are not distributed evenly over the globe; they are concentrated in a few countries, and, therefore, their impact — concentrated on small areas of the Earth's crust—can be of considerable local importance.*

* The existing trend towards more and heavier equipment and tractors is demonstrated by the fact that Brazil ordered, for 1976 alone, 70,000 tractors and 10,000 harvesters (*Newsweek*, 21 April 1976).

Terracing

One of the most important and widespread metapedological techniques is reducing the slope of the cultivated land. The general term "terracing" embraces different types of slope management, including the terracing of steep mountain areas of the Mediterranean or Himalayan type, the "rideau" or "lynchet" fields of Western Europe, the paddy fields of Eastern Asia, and others.

A general definition of terracing has been given by Spencer and Hale (1961):

> A terrace (in agricultural terminology) is any artificially flattened surface on which crops are growing subsequent to the flattening, no matter how crude, how small or how purposeful. The placement of any obstruction laterally across a slope will induce accumulation of mineral and organic soil material behind it, resulting in flattening of a portion of the surface of the slope.

Terracing—or land leveling in general (Yaalon and Yaron, 1966)—activates various pedogenetic processes. Lithosols will tend to convert into a true soil of the terra rossa or brown rendzina type, or even a latosol type, depending on the geographical properties of the area. As the total soil loss from a slope is proportional to the 1.6th power of slope length (Smith and Wischmeier, 1962), a clear decrease of soil loss results from the mere fact of having converted a continuous slope into a series

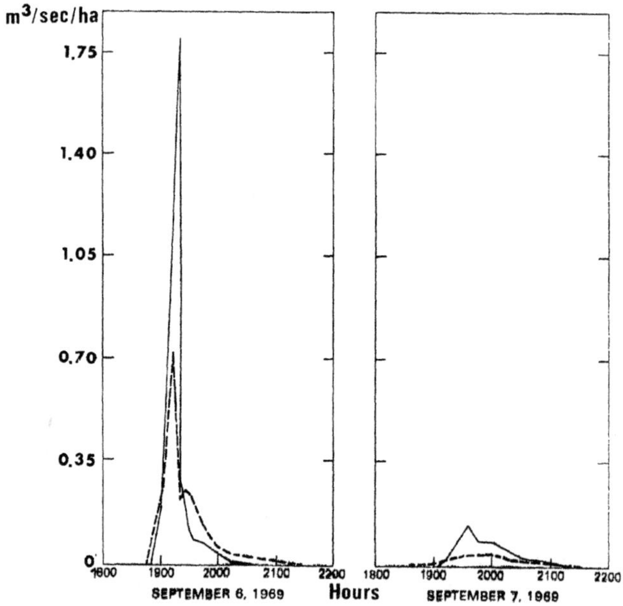

FIG. 11. Effects of terracing on surface runoff. Full line — hydrograph of unterraced land; dashed line — hydrograph of terraced surface (Curtis, 1971)

FIG. 12. Two terraced surfaces prepared for new vineyards, in central Kaiserstuhl, forming part of an 18.5-km^2 project; Germany, 1968 (Haserodt, 1971)

of short slopes or even steps; the shorter the slopes—i.e., the greater the number of terraces on that slope—the more soil erosion will be hindered.

Another important aspect of terracing is the improved infiltration of water into the soil (Fournier, 1972). Terracing will reduce and limit the extent of any gullies formed, preventing attainment of runoff velocities causing erosion. The difference in volume of runoff on terraced and unterraced land is demonstrated in Figure 11; the peak flood from terraced land is only half of that of unterraced land, and only one-third to one-half in terms of volume.

Terracing—of all types—is estimated to extend to over 0.5% of all agricultural land (Louis, 1960); its distribution of terraces of all the types does not correspond to the distribution of agriculturally problematic lands (Spencer and Hale, 1961). Construction of terraces seems, therefore, to have been motivated more by cultural tradition than by environmental considerations. Terraces exist even where agriculture would be possible in their absence, albeit at lower yields (Fig. 12). On the other hand, terraces may constitute the only arable terrain in their area. One may say that terraces demonstrate a permanent impact of Man on the surface of the Earth.

Different types of terraces are found in many climatic regions of the world (Fig. 13).

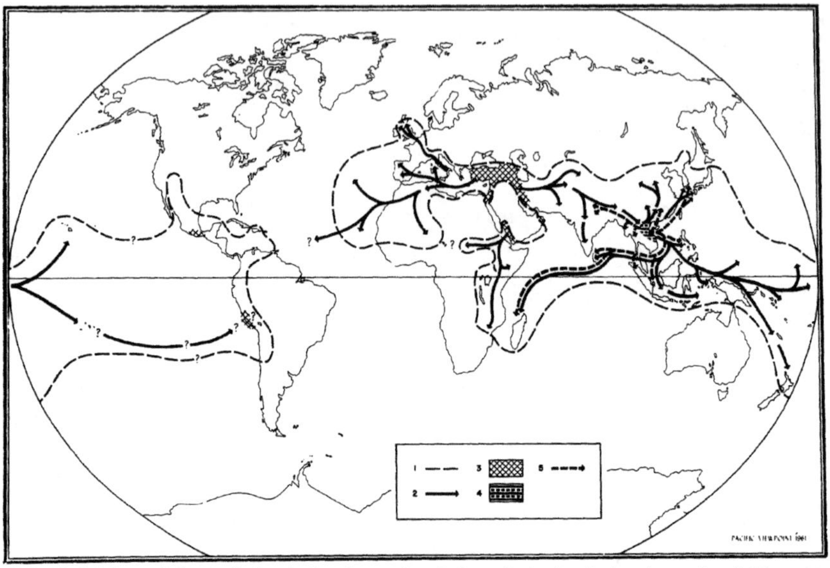

FIG. 13. Origins and dispersal of agricultural terracing. 1. Outer limit of agricultural terracing. 2. Dispersal of earliest terracing technologies. 3. Region of origin of earliest terracing technologies. 4. Region of origin of wet-field terracing. 5. Dispersal of wet-field terracing (Spencer and Hale, 1961)

Spencer and Hale (1961) define four major types: the lynchet-row and furrow-rideau type of Northwestern Europe; the steep-slope terrace of the Mediterranean-Himalayan-Andian type; the channel-bottom weir terrace; and the wet-field terrace of Southeast Asia.

Lynchet-row and furrow-rideau terraces

The improvement of natural land forms is so old a custom that it is practically impossible to distinguish between the "original" land surface and the "managed" one. Hempel (1971; Fig. 14) introduces seven different types of changes in the relief as practiced in Norway, without any representing a "genuine" terrace; other European countries would undoubtedly furnish further examples. These seven types demonstrate that leveling is practiced on slight slopes, whereas intermediate and steep slopes tend to get converted into terraces.

Fields of *lynchets*—neighboring fields limited by a steep slope—are quite common in different parts of Britain, for example in Yorkshire (Smith, 1975). Typical examples are situated near Appletreewick (Figs. 15, 16), where strip lynchets are composed of soil containing boulders and stones removed in the clearing stage. The change induced in the natural slope and the structure of the lynchet clearly indicate the human origin. The breaks in the slope have clearly prevented erosion of the sandy loam overlying the reef limestone.

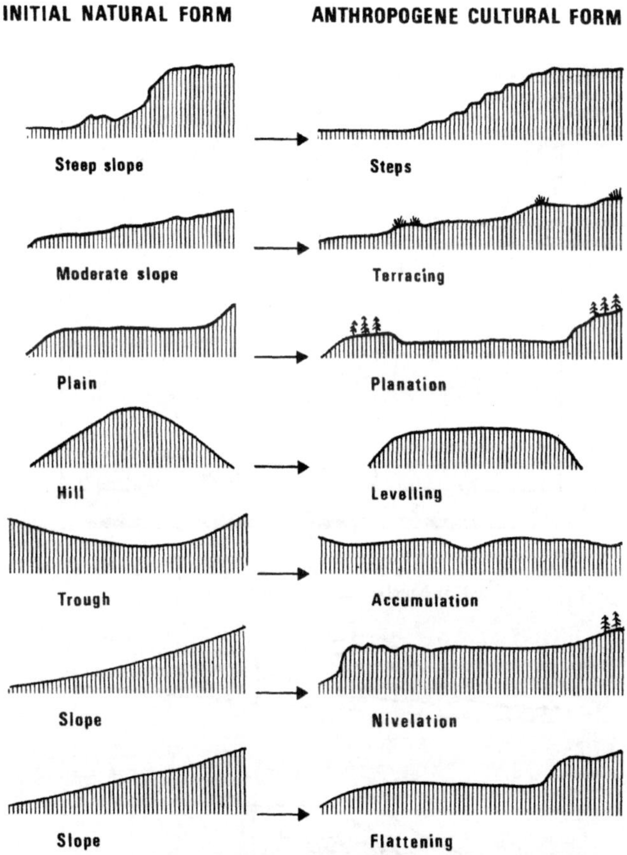

FIG. 14. Natural (left) and anthropic (right) relief in Norway (Hempel, 1971)

The related *ridge and furrow* technique exists in the Midlands, especially in Buckinghamshire (Mead, 1954, Fig. 17). This type of ploughing ridges, varying from scarcely discernible to over 90 cm high with furrows up to 13 meters wide, may have been intended as drainage. Removal of up to 620 m^3/hectare has been involved, a feat of no mean proportions particularly when considering the available tools. In North Buckinghamshire, 28 % of the shire, amounting to 3,500 hectares, was thus treated. Some 3.5 million tons of earth were lifted here over an average of half a meter.

In northern France, particularly in Picardy, the landscape manifests terrace-like features, slopes separated by leveled surfaces, named *rideaux*. They are defined (Aufrère, 1929) as slopes of steep inclination, generally rectilinear, limited by plane surfaces (Fig. 18). The upper plane of the rideau results from cumulative tilling of the soil. Aufrère, after lengthy discussion, considers these phenomena as artificial,

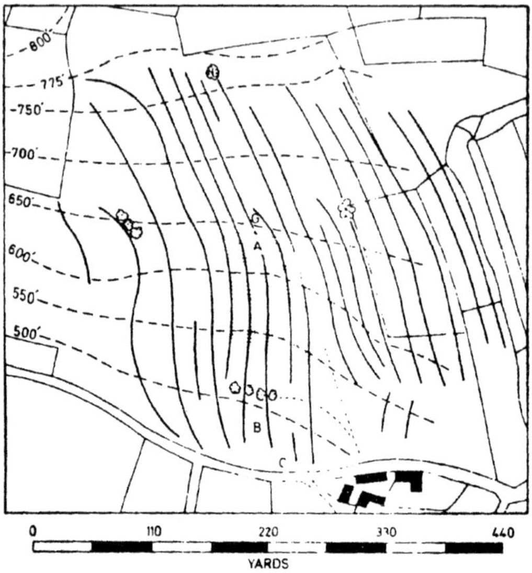

FIG. 15. Plan of lynchet hill slopes at Appletreewick (Smith, 1975). Numbers at the left: elevation in ft.

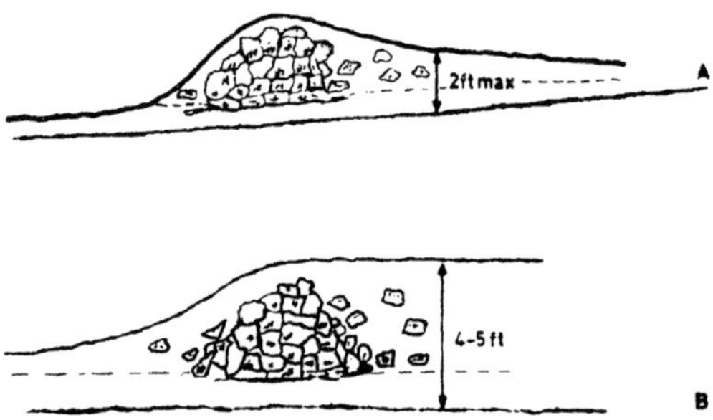

FIG. 16. Variations in the lynchet riser form at Appletreewick (Smith, 1975)

in a few cases perhaps based on an older natural situation. Sion (1940) holds a similar view, and considers the rideaux as a preventive measure against soil sliding and creeping on medium slopes. Sion even sees, in the French Midi, a transition between the genuine rideaux and the Mediterranean type of terracing, where a small wall of stones replaces the earthen slope.

Rideaux have been found in Picardy, Artois, near Nancy, in the Vosges, in the Basin of Paris, and even in Andalusia (Humbert, 1975). They correspond to the lynchet fields in Dorset, Wiltshire, and Buckinghamshire. The lithological background of northern France and several English shires is quite similar, consisting of

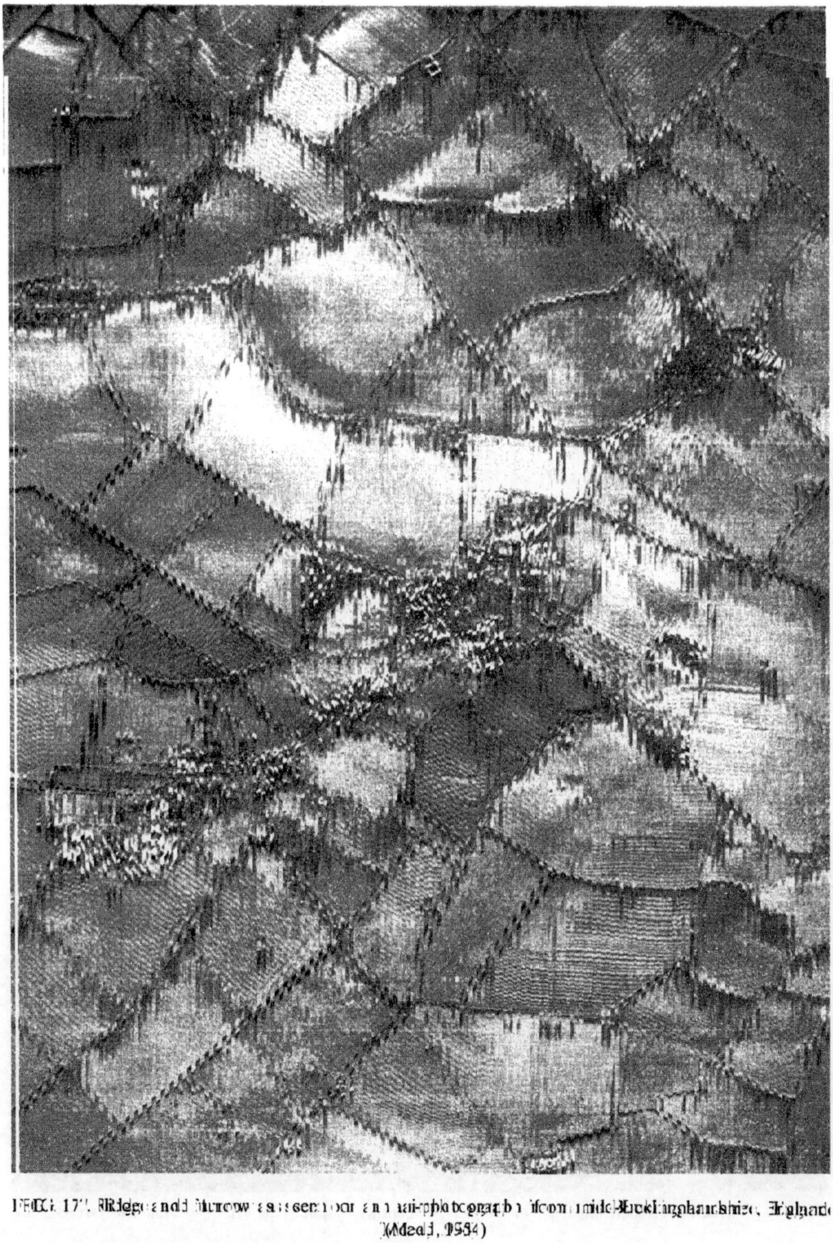

FIG. 17. Ridge and furrow as seen on an air-photograph from mid-Buckinghamshire, England (Mead, 1954)

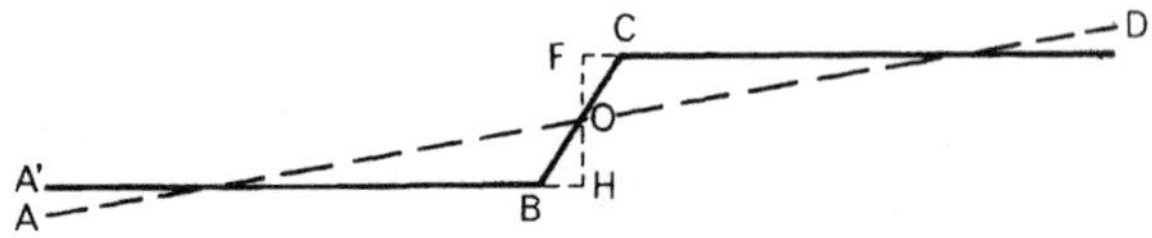

FIG. 18. The structure of a *rideau*. AD = the original surface. AB = horizontal surface limited by the slope of BC. OF = OH (the depth of the cutting equals to the height to the added material). After Aufrère (1929)

chalk, limestone, and clays (limon). Similar types of terraced fields can be found near Dresden in Germany. Demek (1969) mentions north Moravia in Czechoslovakia, where some 40% of the district consists of terraces similar to the rideaux-lynchet type.

Mountain terraces

Those terraces in steep mountainous areas, characterized by vertical, stone-built walls, may be considered the most genuine ones, as they represent the major distinct, man-made, important antierosive agent. In some mountainous regions, as in Nepal, where the soil is deep and stone is not available, the vertical walls are fortified by vegetation only (Fig. 19). The rule, however, is walls built of stone, loose enough to allow water to percolate from the earth fill (Fig. 20). The width of the terraces decreases with the steepness of the slope.

FIG. 19. Terracing of medium-high mountains in central Nepal. The walls holding the terraces are stoneless

FIG. 20. Mediterranean type of terracing Batir, near Jerusalem: terraces built in middle-upper cenomanian limestones and marls

Terracing dates to very ancient times, and was employed around the Mediterranean in the Classic period to an even greater extent than today. A depopulation of the mountainous districts of the most of the Mediterranean countries (southern France, Italy, the Maghreb) has led to abandonment, neglect, and destruction of the terraces, with subsequent accelerated soil erosion. The present trend of population movement away from the mountains strengthens this process.

The Andean culture succeeded, by very elaborate terracing, in converting even the steepest slopes from forest to agricultural use (Bennett, 1939). The stone-retaining walls, supporting these terraces, were usually up to five meters high, reaching in some instances 15 meters (Fig. 21). The high standard of construction employed would permit even at present cultivation of the fields.

Channel bottom terraces

A channel bottom terrace is one of the very simplest forms of terracing (Spencer and Hale, 1961). It consists of a simple weir built across the bed of a shallow, nonperrenial water course which does not carry a large volume of water. This method, practiced today by the semisedentary population of semiarid and arid areas (the Maghreb, Sinai, and others), has its origin in times immemorial. Little labor was invested in making the field, consisting of the naturally flooded area bordering on the

FIG. 21. Terraces in Machu Picchu, Peru. Stone-built walls from pre-Inca periods are preserved rather well in the wet, hot equatorial climate

channel, into a plane surface. Even if the original intention had been to spread the water over the crops only, the weirs acted, in effect, to retain the sediments carried by the floods (Kedar, 1957a, b). Accumulation necessitated increasing the weir's height, once a true plane surface was attained.

Wet field terraces

Wet field terraces are distinguished from all previous terraces by having absolutely horizontal surfaces to permit irrigation with as little water as feasible. If carefully established, only a few cm of water will be required to cover the whole field: an economic, as well as an erosion-controlling advantage. This type of terracing, endemic to Southeast Asia, puts moderate as well as steep slopes to good use. These wet terraces with their hydromorphic water regime and reduced oxidation convert the original soils (alluvial or latosoils) into gleys (Yaalon and Yaron, 1966).

Figure 13 shows the dispersion of agricultural terracing throughout the world. The extent of arable land achieved by this practice is considerable, especially in the Old World: practically all of South Asia practices terracing of the last three types; East and North Africa practices that of the second and third types; Western Europe and the Mediterranean employ the first two types, whereas in the Andean area it is mainly the second type of terracing that is found.

THE GEOMORPHOLOGICAL EFFECTS OF IRRIGATION

The Extent of Irrigation

In 1976, 230,556,000 hectares were under permanent irrigation in 118 countries (FAO, 1977). China alone contains over a third, and Asia (excluding the U.S.S.R.) 78% of the world's total. With an average irrigation of 10,000 m^3/hectare (a reasonable value in semiarid areas), total irrigation requires some 2,306 $\times 10^9$ m^3 yearly of the different water resources (rivers, springs, subsurface water table). Extraction adversely and strongly influences the height of the water table of the area from which the water is being taken. Moreover, irrigation is by no means harmless in all cases: with water of a high sodium content, the soil becomes structurally degraded, its permeability decreases, and even fertile chestnut soil may turn into solonetz (Yaalon and Yaron, 1966).

Already in ancient Egypt, masonry dams to control the Nile and provide irrigation water were constructed (3100 B.C.). Later, in 2200 B.C., Hammurabbi in Babylonia issued regulations concerning the maintenance and operation of irrigation ditches. In the seventh century B.C., Sennacherib in Assyria built an 80-km-long stone canal to bring water to Nineveh for irrigation purposes. In ancient (and even in modern) countries, a clear distinction between "canal" and "river" is frequently meaningless and even impossible (Jacobsen and Adams, 1958). If the canals are large and left without man's control for a sufficient period of time, they will develop a "natural" regime through meandering, cutting of banks, silting-up, etc. Continuous maintenance is needed to keep these channels in a proper state (p. 89). It is believed that the practice of irrigation spread from Babylonia to China (Sherlock, 1922). By 200 B.C., the Tu-Kiang dam provided water for 200,000 hectares; between 589 and 618 A.D., the 1,100-km Imperial Canal in China was built. Pre-Columbian irrigation was practiced in Mexico, Peru, and other South-American countries.

These examples may suffice to illustrate the fact that artificial irrigation is as old as history itself, and its environmental effects are an integral part of the processes shaping the landscape.

Salination Caused by Irrigation

Accumulation of salts results from extended irrigation, especially if the water covers the ground for some time, with accompanying serious effects.

A documented case of salt accumulation in the soil caused by continuous irrigation is that of ancient Mesopotamia (Jacobsen and Adams, 1958). Today's soil salinity in Mesopotamia is only due in small part to marine influences; the major cause has been that of salts leached from sedimentary rocks and added with the irrigation water. As the water evaporates, the calcium and magnesium precipitate as carbonates, leaving the sodium ions dominating the soil solution. Unless they are washed down into the groundwater, the sodium ions are absorbed by colloidal clay particles, deflocculating them, and causing the resulting structureless soil to become impermeable to water. Ancient control of the water table was based carefully on avoidance of overirrigation and on the practice of letting the land lie fallow in alternate years.

The earliest recorded salination in Mesopotamia occurred in 2400–1700 B.C. (Eckholm, 1976). Fields, which in 2400 B.C. were still noted as salt-free, according to an archive from 2100 B.C. developed sporadic salinity resulting from 300 years of exploitation. Wheat, sensitive to salinity, was replaced by barley. Where wheat still grew, its yield went steadily down from a relative value of 100 in the year 2400 B.C. to 50 in 2100, and to 33 only to 1700 B.C. (Jacobsen and Adams, 1958). Cases of salination are, however, not limited to ancient periods. Witness Pakistan, where—as recently as 1962—severe salinity covered about 2.6 million hectares, increasing by 20,000–40,000 hectares per year (Simons, 1967). In the area of upper Sind, with 1.5 million under canal irrigation, 400,000 hectares have a salt content higher than 0.4%, prohibiting cultivation. In the lower Sind area things are even worse: the Ghulam Muhammad Barage canal system covers 800,000 hectares out of which 526,000 hectares are uncultivable for the same reason.

The remedy for excessive salination is, at least partly, leaching the saline soils combined with drainage. It stands to reason that if a part only of the efforts invested in grandiose irrigation projects were turned to less grandiose but extremely important drainage purposes, the problem of salination would be decreased.

Subsidence of Soil as a Result of Water Extraction

Whereas introducing excessive quantities of water will result in salination, excessive pumping of water will—under certain conditions of soil structure—cause the soil to shrink and subside (Poland and Davis, 1956). One of the best-known regions where this phenomenon was studied is the area of Los Banos—Kettleman City in California (Bull and Miller, 1975). The area is part of the agriculturally developed San Joaquin Valley. Water for irrigation has been pumped from wells at increasing rates, from 456 million m^3 in 1945 to 1,480 million m^3 in 1953. Compare these volumes with that of 1924, a mere 42 million m^3; the rise was one of more than 34 times in 30 years. Following the 1953 peak, the volume drawn has stabilized at an average annual level of 1,233 million m^3.

The water has been pumped from alluvial fans of the Quarternary Age. These consist of clays, silt, and sand to a depth of 270 meters. An intermediate layer of clay of 15 to 40 meters seals the lowe water table, situated a hundred meters below the surface, and from which most of the water has been pumped. This lower water table drops 2.5 to 10 m per year.

The affected area extends some 565,500 hectares (39 by 145 km) to a subsidence depth of from 1 to 3 meters, increasing in places to up to 9 meters. This has been accompanied by compaction of the irrigated areas as a result of the unconsolidated parent material of deltaic, lacustrine, and alluvial origin being leached out. A similar chain of events has been reported by Rosepiler and Reilinger (1977). In the Pecos, Texas, area, some 150,000 hectares subsided vertically well over 20 cm in the period 1934–1956, with the water level going down as much as 60 meters over the same period.

Changes in Sediment Transportation and in Rivers' Regime

The influence of drawing water from rivers can be considerable even at a distance downstream. If we return to the case of Mesopotamia, it has been postulated (Davis, 1956) that the changes over the last 4 to 5 millenia in the coastline of the Persian Gulf are, at least partly, a result of irrigation methods which induced accelerated sedimentation at the outlet of the rivers. This coastline has advanced over the last 2,700 years by more than 200 km (Davis, 1956). Until 700 B.C. the Tigris had a separate outlet in the Gulf; today, the river uniting the Tigris and the Euphrates (Shat-el-Arab) is 180 km long. The present extensive delta has an area of more than 7,000 km^2. Withdrawal of water for irrigation may change the regime of a river by keeping its bed dry during certain periods of the year. This has happened to the Marica River in Bulgaria (Ziapkov, 1974), used for irrigation since the 14th century. From 1868 to 1872, irrigation canals were built, serving some 2,500 hectares. Since the end of World War II, a new irrigation scheme of 700 km of canals, supplying 22,700 hectares, was added, necessitating hundreds of small dams and reservoirs on the reaches of the river's tributaries. As a result, summer discharge of the Marica River has decreased to a small fraction of its annual average, between 1.8 and 3.5 %.

Decreased water discharge by a river may significantly lower the level of lakes fed by them. The effect has been calculated for a case where irrigation began only 150 years ago, and where data and observations exist. It concerns the region around the Great Salt Lake in Utah, where the irrigation water is taken from a closed basin, and natural water loss is by evaporation. The use of water for irrigation rose (in millions of m^3) from 247 in 1850 to 1,234 in 1900 and to 1,604 in 1950 (Whitaker, 1971). By taking the past volumes of water extracted thus into account, the hypothetical level of the Lake could be calculated and compared with actual level (Fig. 22). The mean difference between the actual level of the Lake and the hypothetic one was 1.7 m in the late 1920s, and much less in the 19th century; today it is some 1.6 m.

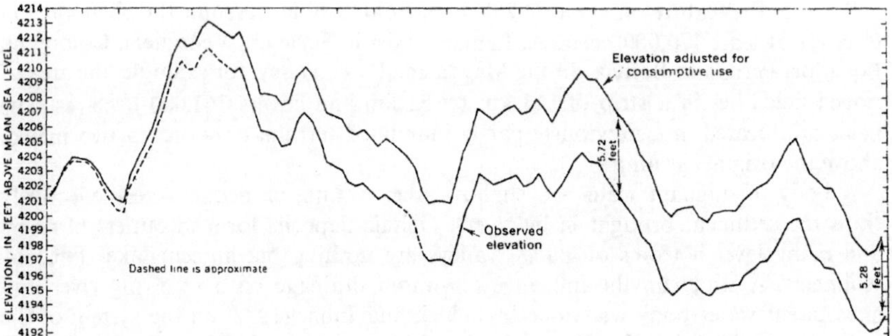

FIG. 22. The effect of water, used for irrigation, on the level of the Great Salt Lake, Utah (Whitaker, 1971)

EFFECTS OF DRAINAGE ON GEOMORPHOLOGICAL PROCESSES

Just as irrigation can change dry land into a quasipermanent water surface, drainage can convert a swamp or lake into dry land.* The effect on the soil properties may be favorable. The lowered water table improves oxidation in the soil (Yaalon and Yaron, 1966) and the soil structure is reinforced. This and the increased rate of permeability favors agricultural purposes. Hydromorphic soils, pseudogleys, etc., formed under the previous conditions of steady state, develop into pseudobrown soils. On the other hand, a complete drainage may trigger off new problems, as was the case of the Hula Lake (swamp) in Israel in 1956 (Karmon, 1960). The exposed peat soils became increasingly oxidized, with resultant autoinflammation of these layers. Only a controlled and steady water supply permitted management of the new situation; a precarious new equilibrium between the reclaimed soil and its water content is now being kept (Figs. 23, 24).

The practice of drainage is certainly not limited to recent times. One of the most striking examples of ancient reclamation is that of wet, hot lands in pre-Columbian South and Central America (Parsons and Bowen, 1966; Denevan, 1970). Denevan sees in the need to establish the so-called ridged fields a proof of the growing density of population in pre-Columbian America. Whereas usually irrigation and terracing of mountain slopes were considered a demonstration of agricultural expansion, existence of cultivated, partially flooded land has been mostly ignored, mainly because such structures seldom survived into the period of European conquests. The ridged fields that have been found in the seasonally flooded plains are a genuine South-American feature. Terrain was reclaimed by improving the banks of rivers, to yield a diversified pattern of terra firma (Fig. 25): built soil platforms surrounded by permanent water bodies; ridged, platformed, or mounded fields in seasonally waterlogged or flooded areas; ditched fields, mainly for subsoil drainage; fields diked to keep water out (Figs. 26, 27).

Some of the features are at least 2,000 years old. Scholars assume the phenomenon to cover at least 170,000 hectares. Remains exist in Surinam, Venezuela, Colombia, Equador, Peru, and Bolivia. In the Magdalena River Valley, for example, the area of ridged fields lies in a strip of 113 km by 32 km and covers 361,000 hectares. The fields are located in a previously partly inundated terrain, only one to two meters above the original swamp.

A body of stagnant water—a shallow lake, swamp, or marsh—very effectively traps the sediment brought in by rivers. Deltaic deposits form in outlets of rivers and many level bottoms of glacial valleys are nothing but ancient lakes filled by sediments. A study on the influence of swamp drainage on a receiving river and subsequent water body was done by Schick and Inbar (1972) on the system of the (swampy) Hula Lake, the Upper Jordan River, and the Lake of Galilee (Israel). The sediment contribution of the Jordan River to the Lake of Galilee is estimated

* Drainage of sea shores is treated on pp. 108–111.

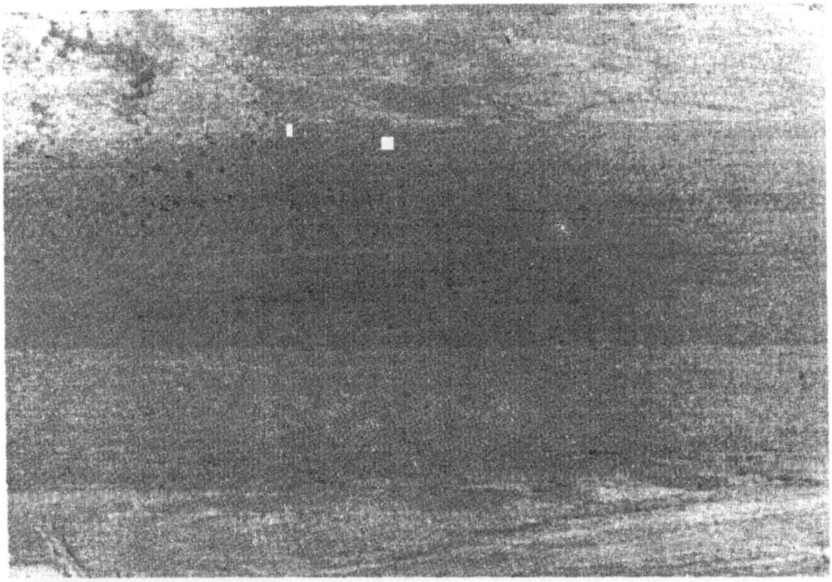

FIGS. 23, 24. Changing of hydrological basins. Above, Hula Swamps before 1956; below, the same area after reclamation. Black peat soils alternate with white lacustrine ones; the streams are regulated and form two channels

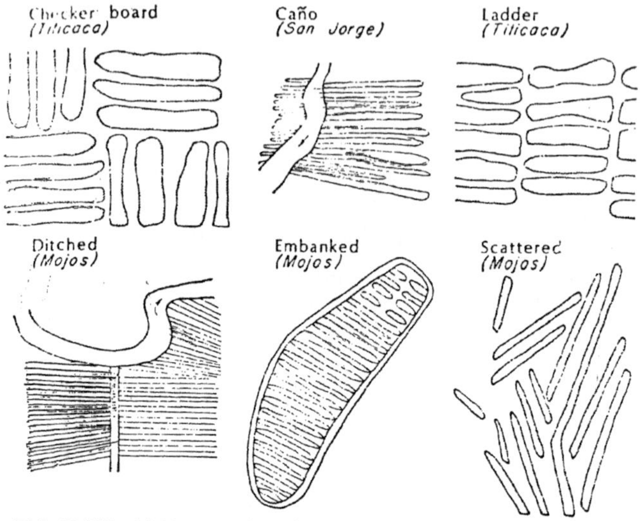

FIG. 25. Ridged field patterns in drained flood plains of South American rivers (Denevan, 1970)

to cause a sedimentation rate of 0.5–1 cm/year over the entire lake bottom. As the lake's exit is closed by locks, it serves as a complete trap. This rate exceeds by a factor of 2.5–5 the value of 0.2 cm/year, found by palynological methods for the past 18,000 years. The higher actual rate is attributed to the drainage of the Hula Lake in 1956/57 (Figs. 23, 24), which previously served as an intermediate sediment trap. Since its virtual disappearance these sediments reach the Lake of Galilee.

EROSION OF TILLED SOIL

"Normal" versus "Accelerated" Erosion

Agriculture, by tilling the soil, removes the protective soil cover and is a potential cause of *accelerated* erosion — as opposed to *natural* erosion. Bennett (1939) estimated the induced erosion to reach 83.55 tons/hectare/year in the U.S. on clean tilled soil, and 0.85 tons/hectare/year only on densely covered soil. These estimates were based on over 50,000 measurements in 13 states, spread out over between 3 and 11 years. Leopold (1968) found the sediment yield of a square kilometer to increase from 28 to 120 tons/year, when the percentage of area covered by crops increased from 10 to 50%.

These erosion rates could mislead as to their accuracy, being average values only. Rather, a well-based judgment on rates of erosion will depend on a number of factors. A main factor is the type of crop. For the same type of soil, Bennett found the following rates of erosion (in tons/hectare/year) for the following crops: alfalfa, 0.62; rotation

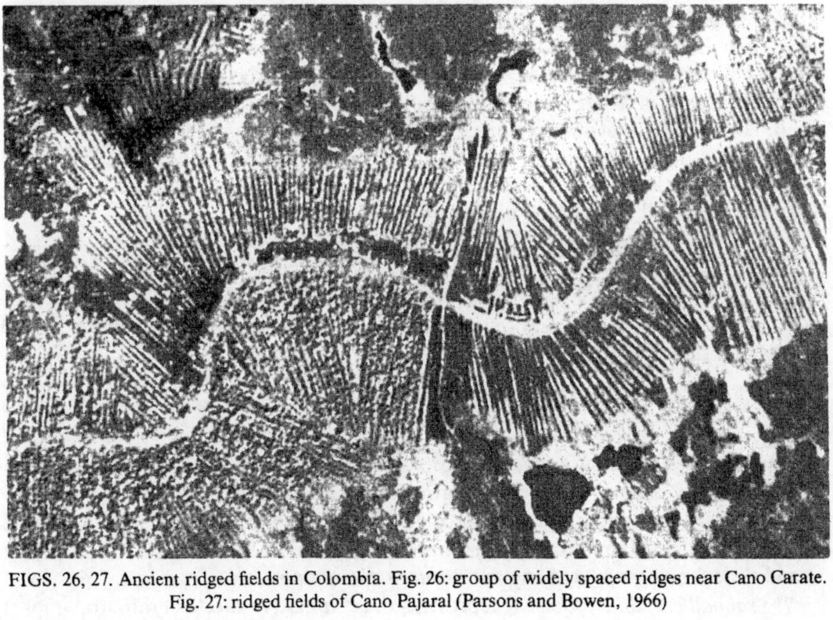

FIGS. 26, 27. Ancient ridged fields in Colombia. Fig. 26: group of widely spaced ridges near Cano Carate. Fig. 27: ridged fields of Cano Pajaral (Parsons and Bowen, 1966)

fields (wheat, root crop, barley), 28.15; maize, 180.31; and open fallow, 170.29. Of the total eroded material carried by the Mississippi, 450 million tons/year (Collier, 1972), some 125 million tons/year are attributed to "accelerated" erosion. Thus, accelerated erosion is defined as the result of an *artificial disturbance of the vegetation cover and ground equilibrium chiefly by the instrumentality of Man and his tools.*

Soil erosion caused by tilling is illustrated by a case study in the Missouri watershed (Bennett, 1928). The amount of annual erosion from the area of one acre, having an average slope of 3.6%, was determined. When ploughed 10 cm deep, the erosion totalled 41.2 tons, and 69% of the rainfall was retained; when grass-covered, however, the erosion reached 0.3 tons only, and 88% of the rainfall was retained by the vegetation cover.

Differences in texture of alluvial deposits distinguish between "natural" and "accelerated" erosion (Strahler, 1956). A change to coarser texture is evidence of accelerated erosion by increased runoff; there occurs, subsequently, an increase in the transporting capacity and in the competence of the stream.

Soil erosion removes, as shown above, also the soluble phosphorus, potash, nitrogen, and lime, whereby the soil's nutritive and economic value is diminished. Whereas, however, nutrients are renewable—although at considerable expense, at times—the soil, once removed, cannot be replaced. Fertilization is, of course, very important: one ton of wheat extracts annually 26 kg of nitrogen, 13 kg of P_2O_5, 16 kg of K_2O and 10 kg of lime. If not reintroduced through fertilizers, the deficiency will, in addition to the evident shortage, also loosen the soil texture, making it more prone to erosion.

Not all soil loss results from tilling only; accessory activities—roads, buildings, etc.—may cause heavy erosion. Bennett (1928) reports about the Patterson Hills County in the Atlantic Coastal Plain, where a giant gully developed over a few years from the runoff of a barn roof only. Gullies of enormous dimensions—30 meters deep, 60 to 150 meters wide—spread out over large areas are perhaps the most conspicuous demonstration of soil erosion in an agricultural landscape. Worn-out agricultural areas seem not to have resulted from extensive cropping but rather to have lost their soil by washing away.

The Soil-Loss Equation

The universal soil-loss equation as expressed by Smith and Wischmeier (1962) and elaborated by Hay et al. (1970), is

$$A = R \cdot K \cdot L \cdot S \cdot C \cdot P$$

where A = loss (in tons/acre), R = rainfall erodibility index (rain intensity), K = soil erodibility factor, L = slope length factor, S = degree of slope factor, C = cropping management factor, P = erosion control practice factor. R and K cannot be influenced by Man; human intervention is responsible for P and C, as well as for L and S, depending on terracing and contour ploughing.

The rainfall erodibility factor R represents the cumulative effects of raindrop impact at maximum rainfall intensity. This factor can be expressed as a function of rainfall intensity.

The soil erodibility factor K represents the inherent erodibility of the soil and is determined experimentally as the rate of erosion from a unit plot, 25 m long, of a uniform slope of 9%, kept continuously fallow and tilled up and down the slope.

$L \cdot S$ is the ratio of soil loss per unit area from a given field to that from a unit plot

$$L \cdot S = \gamma^{0.5}(0.00611 + 0.00076\mu^2)$$

where γ = field length and μ = dominant slope in %.

C = *ratio of soil loss* from land cropped under specific conditions to that from continuous clean-tilled fallow land.

P = *parameter* representing the reduction in soil loss resulting from soil conservation practices such as contour tilling, contour strip tilling, terracing, etc.

A different version of the soil-loss equation was developed, from the original Smith and Wischmeier's, by Williams (1974)

$$Y = 11.8 \cdot (Q \cdot q_p)^{0.56} \cdot K \cdot L \cdot S \cdot C \cdot P$$

where Y = sediment yield from an individual storm, in metric tons; Q = storm runoff volume in m^3; q_p = peak runoff rate in m^3/sec. K, L, S, C, and P retain their original meanings. This equation can be used to estimate the effects of land use and conservation practice as obtained on small areas.

Wind Erosion of Tilled Soils

Cultivation of vast surfaces lacking seasonally water in substantial amounts invites wind erosion or deflation (Strahler, 1956). Therefore, soil erosion occurs particularly on the border of aridity. Natural catastrophes such as a period of subnormal rainfall, reinforced by destruction of the vegetative cover through tilling, expose the soil to the effects of wind erosion. The soil turns to a dry, powdery substance (Bennett, 1939), lying loosely on the surface of the land.

Several cases of this anthropogenic wind erosion of soil have been studied in the U.S. and in Canada. Among the better-known is the "Palliser Triangle" in southwest Saskatchewan, a dry farming region with average annual precipitation of 375 mm (Turner, 1955). Recurring droughts in the 1930s combined with intensive land tilling converted this area to a "dust bowl." Another "dust bowl," formed in the same period, is that of the Midwest of the U.S.—Texas, Dakota, and Oklahoma.

Whitfield and Perrin (1939) studied erosion by wind and dune formation in the region of Dalhart Dallam County in Texas. No dunes were known in this county before 1930. As from 1935 dust storms were, however, recorded, numbering from 61 in 1935 to 71 in 1938. The wind was building up dunes, after the surface cover had been destroyed by tilling, grazing, drought, and heat. One particular field of 190 hectares was studied, revealing the following sequence:

1930: area covered by natural grass vegetation.
1931: area cultivated for the first time; sorghum planted in rows of 1 m distance.

1932: planting repeated.
1933: planting repeated.
Because of drought and crop failure, only one crop was harvested during these three years.
1934: area left idle.
1935: *idem.*
In 1935, 57 sand dunes were counted, their height ranging from 0.3 to 3 meters, their length averaging 53 meters, and their width, 37 meters. The subsoil between the dunes was hard and eroded to a depth of 25–30 cm.
1936: intervention through deep ploughing and revegetation.
1937: beginning of grazing.
1938: nearly all the dunes had disappeared.

During the years 1930–1937 many areas of Texas and Oklahoma were covered by so-called "black dusters," extending over areas of 300 to 1,000 km^2 (30,000 to 100,000 hectares; Bennett, 1939). Some dust storms covered areas of 1,300,000 hectares, transporting 7 million tons of dust, or the equivalent of one centimeter of soil.

Not only arid regions can fall prey to wind-caused erosion: the phenomenon may occur even in such humid regions as the fenland around Ely in east Anglia (Arber, 1946), where dust storms became a frequent occurrence during the 1930s. The peat of the fens suffered from wind erosion nearly every spring. The drainage of these naturally waterlogged soils contributed to the dryness of the soil, although there were other contributory factors, such as heavy frosts in preceding winters, low amounts of rainfall in the months February–April, and — especially — changes in the type of crop culture. In earlier years, the principal crop had been wheat, rye, beans, and oats. In 1929, however, there was a change to sugar beet. This crop needs a different soil treatment, especially deep ploughing, and thus the frost was able to penetrate and break the soil structure. As compared with the former crops, more weeding and chopping were needed. Furthermore, beet has less foliage than cereals, exposing more of the ground to the wind. Also, in contradiction to cereals, the roots (= beet) are removed from the soil. The *change in land use* had implications on the steady state between the crop and the soil, as it had been developed, with the disastrous results described.

Tilled Soil and Frost Morphology

The effect of the thawing and freezing cycle on tilled agricultural lands has been largely neglected, in favor of that directed to gully and wind erosion. It should, however, be kept in mind that alternate freezing and thawing of saturated soils assists in their erosion (Ramser, 1917). As this layer of loosened soil thaws,* it settles with a tendency to move slightly down the slope, and creeping ensues. It is very common for heavy rains to occur directly after the thaw, with the result that his loosened soil is washed away. Probably no other combination of natural factors operates more effectively to rob a field of its most fertile soil over any given period of time.**

* On the effects of thawing of permafrost, cf. pp. 80–81.
** On the effects of frost on highway embankments, cf. p. 82.

Some Estimates of the Rate of Soil Erosion

Evidence of agricultural soil erosion appears throughout the world; as Bennett (1939) states, "the history of soil erosion is an integral part of the history of agriculture." It can be traced as early as in ancient Mesopotamia (p. 37). In China, soil erosion was preceded by soil exhaustion. The early home of Chinese civilization, the northwestern loess region, is now a landscape of huge badlands: a weird landscape that resulted from overintensive cultivation. In the Classic Roman world a sudden increase of sedimentation occurred in the second century B.C., coinciding with intense rural activity (Judson, 1968). Cores taken from the Lago di Mentana disclose the rate of erosion, which before 200 B.C. was 2–3 cm/1,000 years, changing later to 20 cm/1,000 years, a rate ten times as high. Even where erosion was never considered a particular problem, as in Germany and France, the 18th century was not without its problems: Vogt (1953) refers to the "Cahiers des doléances" ("Books of complaints") of 1789, in which a changing cultivation was blamed for soil erosion on the slopes of eastern France and southwest Germany. As a result of these complaints, temporary cultivation was prohibited at the end of the 18th century and damaged slopes were reforested. In the eastern U.S. 400 years of practiced agriculture have deeply influenced soil erosion: Nelson (1966) estimates the rate of erosion in the Chemung Valley in the states of Pennsylvania and New York to be 42.5 cm/100 years. Curry (1971) estimates the U.S. average for tilled fields to be of 2 to 3 meters/1,000 years, whereas the preagricultural rates were, by his estimate, 5 to 10 cm/1,000 years only.

SOIL CONSERVATION

Means of soil conservation include erosion control, such as contour ploughing, strip cropping, ridging, and terracing. Ridging and terracing were discussed above; the practice of contour ploughing, even though mentioned, merits a more detailed examination.

Conservation by Contour Ploughing

Contour ploughing has been recommended as a prime conservation means since the late 1920s by Bennett (1928) and Lowdermilk (1927) and has become the standard practice in areas threatened by erosion in the U.S. and elsewhere. The practice is—as said earlier—of use for slopes below 18 %, above which it should, for erosion-preventing as well as for economic reasons, be replaced by terracing. Contour ploughing serves both under medium as well as under high rainfall, as was demonstrated by six years of study at the Mississippi Experimental Station in Holy Spring (Greer, 1971).

The difference between contour ploughing and up-and-down ploughing, in terms of soil erosion and nutrient loss, is considerable, as was shown quantitatively for the Lake Balaton area, Hungary, by Tóth and Fekete (1974). In 1964, soil and nitrogen

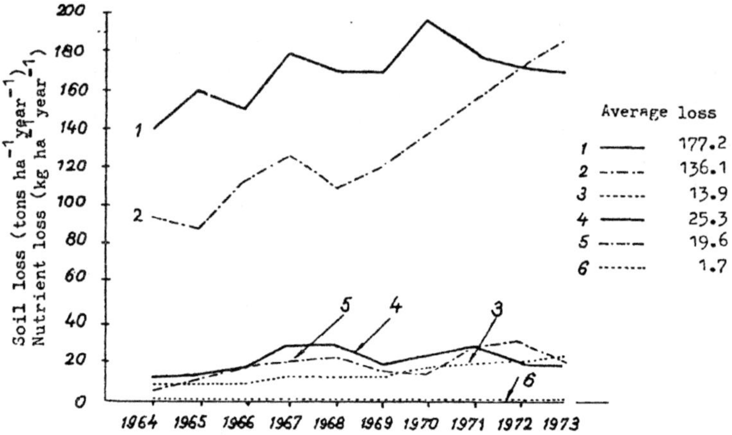

The effect of cultivation on soil erosion around Lake Balaton.
1. Soil loss for down-hill cultivation.
2. Nitrogen loss for down-hill cultivation.
3. P_2O_5 loss for down-hill cultivation.
4. Soil loss for contour cultivation.
5. Nitrogen loss for countour cultivation.
6. P_2O_5 loss for contour cultivation.

FIG. 28. The effect of cultivation of soil erosion around Lake Balaton (Tóth and Fekete, 1974)

loss for up-and-down ploughing were 10 and 17 times, respectively, as high as for contour ploughing; in 1973, the respective values were 10 and 7 times (Fig. 28).

Contour ploughing requires no small degree of craftsmanship, careful use of tractors, and exact leveling measurements. Furthermore, it is highly important that the ploughing does not cut a rill or gully, small as they may be, to prevent erosion perpendicular to the furrows. If this precaution is disregarded, the erosive process, instead of subside, will be intensified, deepening and enlarging the rill or the gully. Although the field itself will not be subjected directly to sheet erosion, the incising gully will advance by headward erosion towards the top of the field.

Developments of this type were studied by Nir and Klein (1974) in a contour-ploughed region near Qiryat Gat, in Israel's southern coastal plain. The soil consists of loess sand; rainfall occurs during 40 to 50 wintery days and averages 350 mm annually. Aerial photos from 1946 and 1966, combined with field studies, revealed the 20-year average incision of four sample gullies to reach 0.21, 0.33, 0.34, and 0.70 m^3/year/meter of gully length, respectively. On the other hand, Seginer (1966) found in the same area the volume of trapped sediment to correspond to that eroded by the gullies. This implies that the slopes, i.e., the fields, were not harmed, justifying the applied contour-conscious practice.

Control of Gullies

Gullies are — the point has been made already — a considerable threat to the arable soil, requiring attention and efforts at control. They develop where the non-concentrated, overland runoff becomes a linear, concentrated one. The highest reaches of gullies are formed on the lower limit of Horton's "zone of no-erosion" (Horton, 1945). By the "waterfall effect" they then advance upwards, at the same time cutting and enlarging their lower channels by increased competence and capacity. Freezing and thawing, if occurring, enhance this process.

The control of gullies cannot be separated from that of the whole hydrographic basin, as otherwise efforts would be directed at *results* rather than *causes* (Jepson, 1939). Building terraces or ploughing contourwise on one area and disregarding adjacent ones will cause any gullies gradually to encroach upon the terraced or ploughed fields and to damage them.

Grass cover is generally considered the best deterrent of gullies, preferably in the form of a grass strip between the tilled fields and untreated areas. It has been shown in the Ukraine (Nikolayenko, 1974) that on a slope of 14°, a 10-meter wide grass strip arrests about 30% of the runoff erosion from a higher-situated sloping tilled field of 100 m length.

Adamson (1974) recommends the following treatment of gullies: filling up by soil, ploughing the adjacent area contourwise, omitting one furrow on each meter (vertically measured), and creating pasture land with the aid of *Lolium rigidum (trifolium subterranea)* by ample fertilization. Results in New South Wales, Australia, over a period of 20 years (1953–1973), are impressive:

	Total annual runoff (mm)	Sediment loss (kg/hectare)
Treated area	167	389
Untreated area	625	36,331

Runoff is reduced by 74% and sediment loss by 99%.

Use of Fertilizers

The tilled agricultural area in the world, 1.49×10^9 hectares, amounted in 1976 to 33% of the total area in agricultural use, the latter including also permanent pastures and meadows (FAO, 1977). Of the tilled area, a major part served to grow cereals, and this part has been growing over the years, from 669×10^6 hectares in 1961 to 838×10^6 in 1977, i.e. at an average annual rate of 1.2%. The increase of the tilled area expresses the growing pressure on the land to produce more food. This pressure could result in accelerated erosion, mainly through exhaustion of the soil nutrients (cf. p. 56). The world's overall consumption of fertilizers over this period grew, however, considerably (*United Nations Statistical Yearbook*, 1977), amounting in N, P_2O_5, and K_2O to 200%, 100%, and 150%, respectively.

To evaluate such impressive progress realistically, values for individual countries rather than worldwide totals should be taken. Some developing countries are still lagging strongly behind when taken on a per hectare or per capita basis. Still, the general trend is not discouraging, and it does not seem at present that an absence of soil nutrients will become a major disaster.

4

Mining and its Implications

INTRODUCTION

Extracting useful material from the Earth's surface is perhaps as old as the use of biological resources: picking fruit, a bird's egg, or a glittering pebble is, after all, one and the same process. More sophisticated use of Earth's materials, such as melting a certain ore to produce a metal, started after the chalcolithic period, and has been around since the fourth or third millenium B.C., depending on the site.

In the Stone Age, each human being used no more than a few kilograms of rock annually for his stone axes. Today, the annual per capita use in the U.S. of rock, sand, and gravel alone, not including ores and metals, is nearly 9 tons (Collier, 1972).

As opposed to agriculture and animal husbandry, mining has always been linked to specific sites, bearing the required ores. Developing technological processes and changing notions on the use of the various materials caused the exploitation to shift, leaving behind abandoned derelict surfaces (Gottmann, 1957). Judging from the dimensions of abandoned sites of earlier exploitation, these activities have had a considerable impact on reshaping the original relief. Significant quantities of slag and iron oxides, resulting from 500 years of operations, have been found in the border counties of England and Wales (Sherlock, 1922). In Mareshfield Parish, Sussex, a heap containing Roman pottery remains, measuring over two hectares and between 0.6 and 3.6 m thick, was uncovered. At Stoke Hill, in the Mendip Hills, a heap of lead working waste was found, measuring 1.6 km by 230 m and between 3.6 and 6 m deep, representing a volume of 380,000 m^3.

The impact of mining on geomorphology consists of changes in the surface, extraction of minerals, their accumulation above the ground, changes in the stability of the upper layers of the earth's crust, and, last but not least, of changes when derelict areas are being reclaimed.

OPEN PIT AND STRIP MINING

Strip mining, on the scale we know it today, dates back some hundred years, when horse-drawn scrapers started replacing the pick and the shovel (Doerr and Guernsey, 1956). Today, heavy equipment is employed to remove unwanted overburden, and often soil of good agricultural quality is incidentally converted into waste. Since open pit mining constitutes about 70% of the world's mineral and rock output (Vita-Finzi, 1974), its environmental impact may be significant. This is particularly so, if after extraction the area is left idle, and stream and gully erosion and sedimentation will occur. Proper treatment and management, on the other hand, can return the area to productive agricultural use (p. 73). The total area involved worldwide in open pit mining is difficult to determine. In the U.S. only, the disturbed area in 1965 was 12,500,000 hectares (U.S. Dept. of the Interior, 1971); one mine, the Bingham Canyon Copper Mine in Utah (Fig. 29), producing 2,445 million tons of ore between 1906 and 1969, involved an area of 5.4 km^2 to a depth of 695 m, the average daily extract being 330,000 tons; this is, possibly, the greatest single open mine in the world (McWhirter and McWhirter, 1975).

New open pit mines amount in the U.S. to 65,000 hectares annually, 80% of which lie in agriculturally exploited areas. That the impact is of a relatively short duration can be seen from the fact, that (in 1956, for example) one-third of the mines had been in operation for less than 5 years, and only one-eighth for over 20 years. The short occupation may, however, be quite intensive. The other country with important strip and open pit mining is the United Kingdom. In East Anglia there are 300,000 locations of extracted material (Vita-Finzi, 1974). Every year, 2,000 hectares are turned over to mineral extracting activity, although a considerable part of this is ultimately reclaimed (p. 73). In the northwest of England, landscape destruction has involved 24,000 hectares in the form of heaps, 24,000 hectares as holes, and 12,000 hectares in other forms (Fairbrother, 1972; see also Fig. 30). Open pit mining is practiced today over the whole world, from the Kiruna (Sweden) iron ore mines in the north to Zaire's copper mines on the equator, from the limestone quarries of the Mediterranean to the gold-bearing glacial drift of south New Zealand. In spite of the difficulties inherent in a complete and detailed description of all events involved everywhere, at least an effort should be made to show and analyze the different means of land use and of the impact on geomorphological processes.

Open Pit Mining of Coal

Of all materials extracted by open pit and strip mining, coal is by far the most important both in the U.S. and in Great Britain. In the U.S., the surface affected by mining (generally) was in 1965 some 1.3 million hectares, of which coal alone involved 41%, as compared with sand and gravel (26%), stone (7%), and iron ore (5%) (U.S. Dept. of the Interior, 1971). Of a total of 14,400 mines, of which 5,900 were for coal, 70% were open pit ones.

In Great Britain, open pit coal mining, although known since the Roman period,

FIG. 29. Bingham Canyon Copper Mine, Utah (*National Geographic Magazine* (1964) **125**, 570)

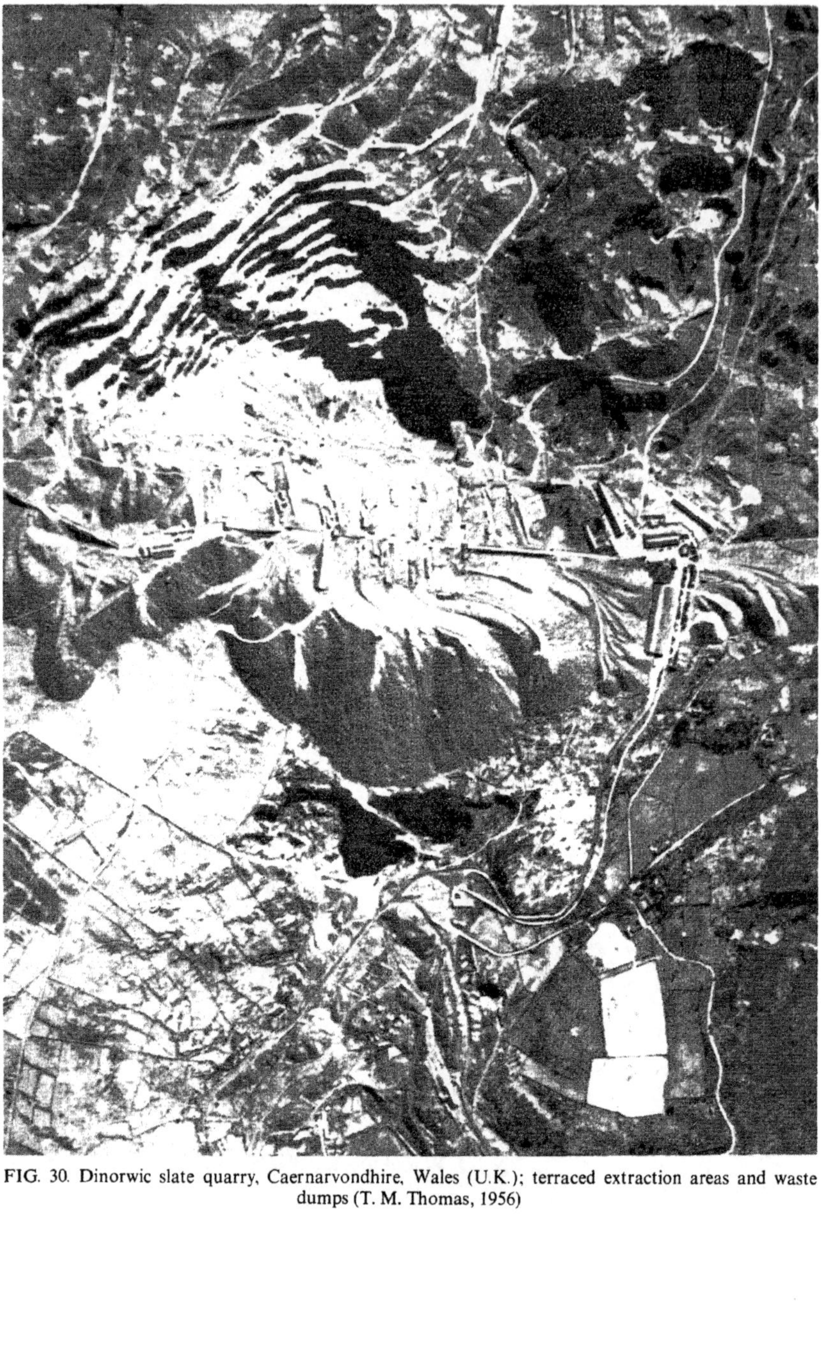

FIG. 30. Dinorwic slate quarry, Caernarvondhire, Wales (U.K.); terraced extraction areas and waste dumps (T. M. Thomas, 1956)

reached its climax in 1941 to meet wartime needs (Davies et al., 1975). Present production is 10 million tons a year. Opening a new stripping mine now involves twenty authorities, to assure that all aspects — environmental future conservation, and reclamation—get proper attention. Thus it became common practice to keep the topsoil for reuse after the mine is exhausted. The long-range impact is thus minimized.

Open pit mining, besides deforming the surface, increases ground acidity and mineralization (Collier et al., 1970) as well as the sediment load of the streams in the stripped area. Since the stripped area consists of loose, man-made detritic material, it is easily attacked by precipitation, frost and thaw actions, as well as by wind, resulting in man-made badlands. A region notorious for such badlands is the Messali range, the iron ore mountains of Wisconsin. The formed pits are 8 km long and 150 m deep, not unlike small "Grand Canyons." Mining the lower grade taconite ore necessitates removal of 30 m cover, giving rise to formation of even greater pits. A taconite mine 8 km long, 300 m wide, and 150 m deep, will create over 570 million m^3 of waste, which, when placed somewhere, creates a new, man-made relief.

The impact of strip mining on sedimentation has been studied in Kentucky (Collier et al., 1970). Strip coal mining in the Beaver Creek Basin began in 1955 and ceased in 1959. The two-and-a-half-year study centered on an area of 180 hectares, comparing two reaches of the river, one in the stripped area which covered 18 hectares, the other in the unstripped one, both reaches having the same hydrographic geometry. Maximum sediment load during storms was 30,000 ppm in the disturbed area and 553 ppm in the undisturbed one. Annual yield was 733 and 9.6 tons/km^2, respectively. The influence of the stripping and mining operations is thus clearly demonstrated.

Effects of Dredging for Washing of Gold-bearing Deposits

Gold appears in alluvia, glacial drifts, and sands, which are dredged and washed to obtain the precious metal. The practice of washing is by huge quantities of water, flushing out all lighter materials, i.e., those other than gold, and leaving the gold on the bottom of the washing pan. This practice creates an entirely artificial flow regime of discharges and velocities probably never attained by a natural stream. The volume of removed alluvia and other clastic material is, no doubt, considerable.

In order to get an idea of the amounts involved, the gold rush on the Southern Island of New Zealand in the years 1857–1960 may be examined, when 815,252 kg of gold were produced (Salmon, 1973). On the basis of data given by different gold mining companies, one ton of deposit yielded 5.1 grams of gold,* which leads to some 160 million tons of material having been removed from their natural setting. These deposits have been redistributed in the lower parts of the gold-bearing rivers. The landscape near Alexandra, Otago, on the Southern Island of New Zealand where considerable mining has taken place, is in fact one great dump of gravel and other flushed-out alluvia.**

* Six companies reported to have produced 1,820 kg gold from 358,750 tons of gold-bearing deposits.

** An additional effect of the gold rush was the deforestation of the hills of south-central Otago: the timber was used for construction of aqueducts for river water, for building, and as fuel. This activity undoubtedly contributed to local soil erosion (cf. p. 19).

Mining of gold by the *placer* technique (Glenn, 1911) in Dahlonega, Georgia, caused entire hillsides to collapse into the rivers as a result of giant artificial streams. Among the largest dredges used, some had a weekly capacity of 98,000 m^3 of gravel, reaching a depth of 15 to 30 m. To produce one million dollar's worth of gold in California, 17 million m^3 had to be processed (Fels, 1935). To this day, thousands of acres along the streams and rivers of California lie derelict, just as they were left 25 or 50 years ago (Beatty, 1966).

The dredging profoundly changes the soil profile of the alluvia: fine sand and silt are being deposited on the bottom and the coarse gravel lands on top, usually in piles. The resulting slope may be 15% steeper than that which existed originally.

THE EFFECTS OF SUBSURFACE MINING

Subsidence of the Surface

Even when actual mining is carried out below ground, considerable above-ground area is occupied by buildings, slag heaps, railways, etc. In Upper Silesia, between 20 and 50% of the area is thus occupied (Demek, 1969), and considerable portions are involved elsewhere as well. A new relief is thus created. The most profound influence of subterranean mining is, however, that causing collapse and subsidence of upper layers.

Taking coal mining in the U.K. only, in the 20th century 16,000 million tons have been unearthed (King et al., 1975), involving subsidence over some 12,000 km^2. Damage to structures has been relatively small, in spite of an average subsidence of one meter.

Subsidence is especially dangerous in the low-lying areas, where floods and drainage problems compound the risks to the local population. Generally, ground movement may affect sewerage installations, highways, and railtracks. Sherlock (1922) describes the phenomenon in Cheshire, where salt has been mined. Nantwich, the area which has yielded 25 million tons in the 18th and 19th centuries, is notorious for the subsidence of its streets and houses. Sherlock quotes some sources, developing a *formula of subsidence*, indicating a "safe" mining depth at which no subsidence on the surface would be caused. The formula of W. Jičinsky (1884, quoted by Sherlock) is

$$S = (T + M) - AT$$

where S = surface subsidence, M = thickness of the coal seam, T = thickness of the rock overlying the coal seam, and A = average coefficient of increase of volume of the coal seams considered as a whole (in several cases $A = 1.01$).

The correctness of the formula and the implied existence of a harmless depth of mining have been contested. Still, Sherlock states (p. 134) that "the amplitude of the subsidence diminishes in proportion as the galleries are deeper below the surface; the diminution is proportional to the depth." He assumes, therefore, that there is no depth of mining totally harmless towards the risks of subsidence.

The Impact of Injecting Liquids into the Ground

Abandoned mines serve increasingly for the disposal of wastes, and where liquids are stored in them, this practice may have marked effects. Liquid waste was injected in a well near Denver, triggering off the first recent measurable earthquake in the state of Colorado (Wilmoth, 1969). Since the disposal of liquids was started, some 1,600 tremors of varying intensity have been recorded in an area where earthquakes were previously practically unknown.

Disastrous effects were also recorded after water had been injected to enhance oil production (Hamilton and Meehan, 1971). In the Inglewood oil field, northwest of Los Angeles, an oil field subsidence of 2 meters has been correlated with the withdrawal of 82.6 million m^3 of oil assisted by water injection. The explanation for these effects lies in the faults of the region, where the injected water reduces the shearing resistance along potential failure plans. The ensuing disastrous collapse of a water reservoir, inundating 160 hectares and damaging hundreds of houses, stresses the dangers inherent in such neotectonic phenomena.

Storing water in abandoned mines may have disastrous results where the water is permitted (or caused) to escape to lower-lying areas, as happened when blasting for road building was carried out below the water level adjacent to a mine in West Virgina that was abandoned 43 years earlier (Wilmoth, 1969). Within minutes a gush of 180 million m^3 water flooded a neighboring town, causing very extensive damage.

Subsidence Above Mines — An Economic Problem?

The question is justified whether calamities such as described above are the inevitable result of mining or whether they could be prevented by proper measures. According to Potts (1975), the hazard of subsidence resulting from subterranean mining can be greatly reduced by proper management of abandoned mines, reinforcement of the roofs of the galeries, and control of their water budget. The ground surface is available to development activities if proper methods are employed, such as building houses on "rafts" reinforced to resist tensional strains, or laying pavement slabs for streets leaving gaps between them. This technology has been used successfully in the new town of Peterlee (U.K.) since 1948.

QUARRYING OF GRAVEL, SAND, AND STONE

Whereas usually the surface effects of underground mining will be limited, it is in the quarrying of sand and gravel that these effects are the most conspicuous. The fact that sand and gravel deposits usually are located in the lower, broad valleys of fertile agricultural land only adds to the impacts of their mining. The scars left by quarrying activities imply efforts for subsequent restoration; such efforts are, however, not always undertaken.

The significance of sand, gravel, and stone stems from their established and general application in building activities, coming only second behind timber; hence these materials are essential for our civilization. Louis (1960) estimated the annual need of every inhabitant to be 1 m^3 of building materials (clay, sand, stone, and gravel), i.e., 2.5 tons. The figure seems to be on the low side if the various development programs (highways, harbors, industrial and domestic building) are taken into account. In fact, available data for some areas are considerably higher, as for south Ontario, where it stood at 17 tons/capita in 1965, or—on the average—for the U.S., some 9 tons/capita in the early 1970s. However, even at 2.5 tons/capita (as a global average), annually over *10 milliard tons* are involved. Quarrying such amounts competes quantity-wise with the combined results of natural erosive forces.

The third important characteristic of sand and gravel is the proximity of their quarries to metropolitan centers, caused by their relative low unit value (Hilpman, 1970) which does not allow high transportation costs (Fig. 31). Pits and quarries seem thus to be an inevitable companion to metropolitan development areas.

The problems connected with this activity have been studied in detail in Great

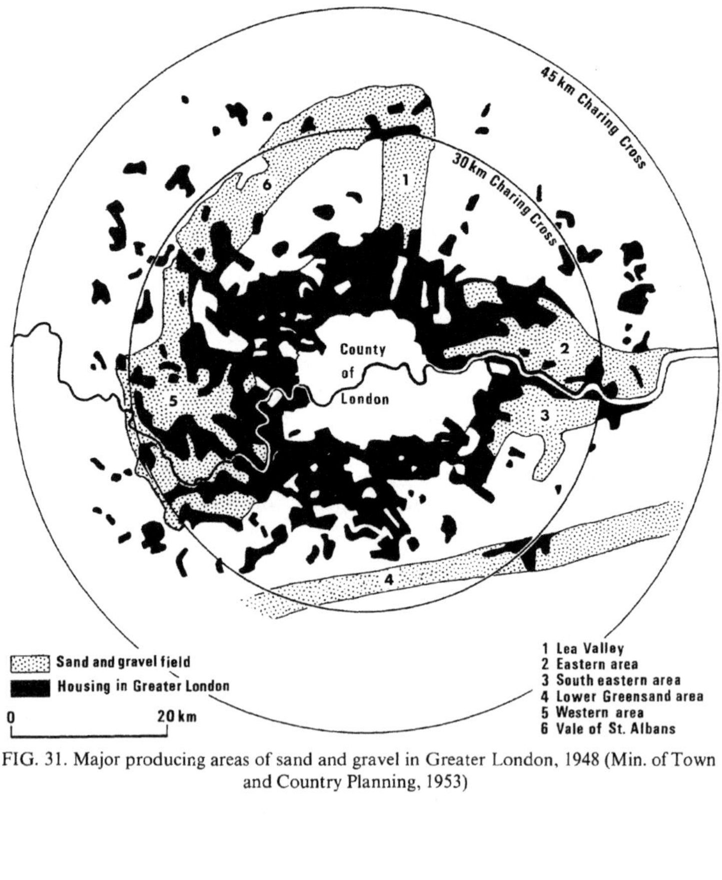

FIG. 31. Major producing areas of sand and gravel in Greater London, 1948 (Min. of Town and Country Planning, 1953)

Britain by a Royal Commission in 1948.* Whereas before 1914 some 1.5 million tons were quarried annually, in 1950 the amount had risen to 23 million tons, affecting (annually) over 600 hectares to an average depth of nearly 4 meters.

Over 1,000 pits were involved (nearly one-fifth of all open pits in the country), most of them in plains with a high water table. This latter fact, sometimes detrimental to initial development of the quarry, was later put to good use in the reclamation efforts of converting spent quarries into artificial recreational lakes, especially welcome because of their adjacency to centers of habitation.

It should be stressed that the extracted material may be nonreplaceable by natural processes. Its origins, in the quarries and pits in the Greater London area, for example, date to postglacial sedimentation, and their complete utilization forms entirely new man-made strata, constituting the top strata with which man will have to live.

MORPHOLOGY OF TIPS AND DUMPS: POTENTIAL INSTABILITY OF ARTIFICIAL SLOPES

The most conspicuous relief resutling from underground mining is one of mounds and heaps of slag and waste, originating from the extraction and processing of the ores. The more precious the ore, the more waste produced. One ton of South African gold ore yields 9.5 grams of the precious metal; in 1975 only, 36 million tons of ore were treated, most of which thus ended up as heaps of waste (Government of South Africa, Mining Survey, 1975). However, exploitation of nonprecious minerals also creates huge amounts of waste, such as 90–95% of the amounts mined for slate (T. M. Thomas, 1956). Sherlock (1922) estimated the world's annual waste production at the beginning of this century to be some 50 million tons. For those 6 minerals involving the largest amounts of waste, we estimate the annual amounts to have reached in 1977 some 4,400 million tons, as follows:

Wastes from mineral production, 1977

Product made*		Products as % of ore**	Waste produced (millions of tons)
variety	tons		
Iron	482,800.000	50	483
Bauxite	79,600.000	55	63
Copper	8,000.000	2	400
Gold	988	0.9×10^{-3}	108
Coal	2,475,948.000	50	2,476
Brown Coal	902,289.000	50	902
Total			4,432

* *United Nations Statistical Yearbook*, 1978 (NY, 1979).
** Vosknil, 1930; Sherlock, 1931, p. 26.

* Ministry of Town and Country Planning, *Report of the Advisory Committee on Sand and Gravel, 1948–1953,* HMSO, London, 1953.

Part of these huge amounts may be used for different purposes, such as road metaling or concrete and brick fabrication. The low unit value precludes, however, utilization outside the immediate neighborhood of the mines. The usual final depository is therefore in the vicinity of the mines, as we may notice near Birmingham and in southern Wales, England, Witwatersrand in South Africa, northern France, the northeastern U.S. and other areas.

The geomorphological activity of the waste mounds is a function of the stability of the slopes, the gradient of which is determined by the interplay of frictional and gravitational forces. As the dumping will be carried out as cheaply as possible, the frictional forces of the usually loose material may be quite low, yielding a precarious equilibrium prone to disrupture by environmental factors. Heavy rains, sudden thaws, earthquakes, or flooding away of the base will create new relations between the different forces, resulting in new slopes (Mulavz and Rybicki, 1977). Toe failure of dumps is sometimes caused by occurrence of plastic clays and the manner of dumping. The failure can expand from the dump outwards at a rate of 6–7 meters/day, and, as a result of this, landslides arise in the dumped waste. The proximity of the dumps to inhabited mining areas then poses acute dangers, of which the most tragic illustration was the disaster at Aberfan. In this Welsh village 2 million tons of waste, heaped over a period of 50 years, in October 1965 suddenly moved and flooded a school to a height of 15 meters, killing 116 children and 20 adults.* The disaster was ultimately attributed to solid movements which redeposited the crest at the bottom, permitting the escape of water previously under pressure, causing a flow triggering off the avalanche. The fact that similar — albeit smaller-scale — occurrences had gone unheeded in the region contributed to the complacency contributing to the disaster.

Occurrences like the one described above are preventable. Tips' management should include minimizing the gradient to below-critical values, leveling of tops, smoothing the slope, and afforestation of old tips. These measures will prevent unpredictable gullies, solid movements, and transportation and unexpected slides. Reclamation of these man-made reliefs should be an integral part of any mining management.

RECLAMATION OF DERELICT LAND

It is now generally agreed in the industrial countries that land made derelict by mining must be reclaimed. Some countries, such as the U.K. and the U.S., have enacted extensive legislation to this end, under which conditions of mining, means to prevent landslides, and terms of restoration of the natural beauty as far as possible are set out.

Costs and Means of Reclamation

Costs of reclamation may vary considerably, depending primarily on what society is prepared to pay for its environment. Some examples may illustrate this. In the

* Keesing's Contemporary Archives, 1965/66, 21683 A, London.

U.K., calculations made in the 1960s indicated reclamation to cost between 8 and 26% of the value of the coal mined. For a more precious commodity, like gold, simple leveling of open pits by bulldozering involved some $6,000/hectare (Beatty, 1966).

Usually no insurmountable technical difficulties are encountered in reclamation. A common way consists of filling back the pits with the original topsoil. Another method is terracing, which efficiently reduces potential erosion of the renewed cover, as shown in Breathitt county, Kentucky, for an 8% gradient furnished with 0.2-m terraces, 2.5 m apart (Curtis, 1971). Peak floods on the terraced plot were 65% less, sediment loss was 52% less, and total runoff was 42% lower than on nonterraced control plots.

The mining wastes being dumped usually attain gradients of 30 to 80°. Considerable regradation, to attain slopes of 25°, is usually required to prevent slides. Draining in case of a high water table and planting of trees or sowing of grass may also be required, as dictated by varying local conditions.

Uses of the Reclaimed Areas

The particular use to which a reclaimed area will be put depends, of course, on its situation, whether it forms part of a mountainous area, lies in flat agricultural land, or just belongs to a developing metropolitan area. Restoration to agricultural use is quite common, provided the proper topsoil has been returned.

Nowadays regional planning is called upon for the general improvement of a derelict county.* Increasingly, reclaimed land—particularly when lying on relatively deep disturbed strata—is used for leisure and recreation purposes. By such processing, highly derelict territories may become preferred recreation areas, of a value exceeding the total value of the extracted ores. Examples include the U.S. Midwest, where strip pits and quarries have been flooded and developed into attractive hunting and fishing areas (Hilpman, 1970). In Illinois alone, two-thirds of the 36,000 hectares of derelict land served (in 1964) recreational purposes (Rosenbery and Klimstra, 1964).

The metropolitan areas of London and of Paris present similar examples. In London, gravel and sand winning affected tens of thousands of hectares. Artificial lakes and sporting centers were created, such as the outstanding Thorpe Waterpark (Hartwright, 1975), consisting of a lake of 120 hectares surrounded by a sports ground of 200 hectares, situated only 30 km away from London.

An impressive case of reclaiming mines is that of the Rheinbraum Co., operating lignite extraction in Aachen-Düsseldorf in West Germany. Since 1945, 44 villages had to be removed because of the mining operations. Subsequent reclamation has now affected 8,900 hectares. Half of the area has been planted with a variety of trees, and in the other half 39 man-made lakes were created. Wildlife preservation and parks are now a main recreational attraction for the population of the nearby large cities. The depopulation, resettlement, and reclamation of the stripped areas cost about 24 million dollars a year, about 5% of the annual sales of the company.

* J. Young, Paradox of the beautiful barren land, *The Times*, 17 March 1976.

5

Means of Transportation

ROADS AND HIGHWAYS

Introduction

The initial clearing of forests provided the areas for settlement and agriculture; the subsequent need for communication between the settlements created the paths and the roads. This type of land use is, therefore, as old as mankind's organized society.

Paths and roads represent major changes on the Earth's surface (Sherlock, 1922). When paved, they act as an armor, protecting the ground against ordinary agents of erosion, simultaneously starting new processes by concentrating rain into a man-made (or road-made) runoff. This is especially the case in semiarid and arid zones, where paths, particularly on slopes, are primary agents of erosion, with a high probability of deflation in the dry season and of gully erosion in the wet one.

Roads, characterized as long and narrow strips of impervious surfaces with concave or convex slopes on their flanks, have been existing in some parts of the world for centuries, and their effects on the geomorphology of their surroundings have been integrated in the processes of erosion and sedimentation. In South America, for example, pre-Inca roads, 7 meters wide and partly paved with bitumen, extended for over 6,400 km, from Quito in the north to Tucuman in the south. The Roman Empire enjoyed at its height over 80,000 km of highways, radiating from Rome in 29 directions to Scotland in the northwest, to Petra in the southeast, to Finisterre in the southwest, and to the Orontes in the northeast (Legget, 1962). Their width was from two to five meters, and their paving material reached up to one meter thickness. Older road networks preceded these ones, such as the Babylonian ones joining the Euphrates with Egypt, employing brick and stone embedded in a bituminous mortar. Most of these roads, especially in Europe, declined with the Roman Empire, and were little used till the end of the 18th century. The 19th century

FIG. 32. Impact of roads on the landscape. Land, previously underneath the Swan River, now part of a road system (Perth, Australia)

saw a new development, road paving with macadam—a layer of compacted crushed gravel; the metaling process with bitumen came as a later improvement. The primitive, the Roman, the post-Roman, and the modern activities thus determine the four periods of geomorphological activity.

That the modern period stands quite apart follows from the very considerable amounts of earth involved, which reached at one interchange in the U.K. 300,000 tons (McWhirter, 1975), as well as from the fact that modern equipment permits road-building on previously water-covered areas (Fig. 32).

Geomorphological Effects of Road Building

The crucial "point of weakness" of a road, from the geomorphological point of view, is the coexistence of a paved, impervious, often inclined surface with the surrounding loose soil. Whether the soil is bare or covered by vegetation is of prime importance for any erosive process. Transferring earth from hills to smooth road gradients will contribute to erosion; landslides are a common highway construction problem (Hayes, 1971). Local climate, and particularly local humidity, will determine whether and at which rate vegetation will cover the new slopes. Where no planting is done, light loess soil will easily form rills, for which an annual rainfall of 400 mm, in two rainy seasons, may suffice (Fig. 33).

FIG. 33. Man-made slope erosion, Qiryat Gat junction, southern Coastal Plain, Israel, 1968. Removing loess for the highway caused these deep rills to form in two rainy seasons only (annual rainfall 400 mm)

On the other hand, a humid climate per se is no guarantee against erosion if the region is one of steep slopes and great rain intensities, such as the monsoon region. Douglas (1971) reports on the effects of road building along the Sungei Gombak River near Kuala Lumpur (Malaysia). The upper part of the catchment consists of the natural forest, disturbed only by the main road from Kuala Lumpur to Padang. A road-widening scheme resulted in intensive soil slipping on the steep slopes, causing a very high suspended sediment load in the river. In the upper part of the watershed, undisturbed by the building activities, the load amounted to 5 mg/liter, reaching in the lower reaches, affected by the works, the high value of 100 mg/liter. The control of road erosion is important where frost occurs commonly, since uncovered slopes are prone to destruction by the cycles of frosting and thawing, even at slopes of less than 2% (Diseker and Richardson, 1961). The subarctic and arctic regions are very susceptible to road construction (Murgill et al., 1976); special techniques will be required when a road should be built on unstable, ice-rich frozen soils. Even more damage is caused by off-road vehicle traffic, as for example in the Denali Highway region in Alaska (Sparrow et al., 1978), especially in poorly drained areas with soils susceptible to erosion. By the construction of roads, runways, houses, etc., the natural thermal equilibrium is upset: the ground becomes isolated from solar radiation, and the natural cover of the ground ("muskeg") is destroyed (cf. pp. 80–81).

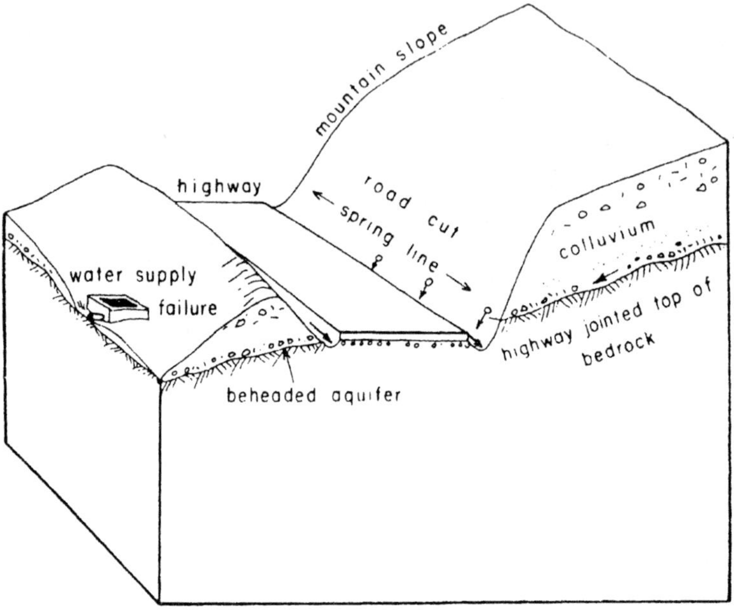

FIG. 34. Influence of cutting a road on the groundwater discharge. Along the spring line, the landslide hazard is enhanced (Parizek, 1971)

The influence of roads on geomorphological processes can be classified as follows: effects on aquifers; changes in the runoff; the "waterfall effect"; siltation of channels; and the effect on and by an environment under permafrost conditions.

Beheading of shallow aquifers

Shallow aquifer systems may exist in many mountainous regions. Cuts into slopes may intercept such groundwater (Parizek, 1971). This will interfere with the natural hydrological equilibrium (Fig. 34). Moreover the slope into which the cut has been made will be prone to landslides as a result of the weakening of its strata. Where cuts are located below a water table, the probability of rock fall and slide increases. Some landslides of this type occurred on the slopes along the mountainous Jerusalem-Tel Aviv highway in the 1960s. The slopes, built of cenomanian limestone over water-containing Motsa marls, began to slide (Fig. 35), requiring the construction of expensive walls to prevent further movement. In this area, only slopes of below 25% remained firm without additional precautions.

Changes in runoff

The roads' impermeability concentrates 90–95% of the runoff, usually through culverts at a few points, to discharge considerable volumes over short periods.

FIG. 35. Man-provoked slide along the Jerusalem highway, 1965. The parent rock consists of cenomanian chalk and limestone. Note the small pine tree dislodged from the woodland

Erosion by the abrupt stream is inevitable since the receiving natural streambeds were formed by gradual, weaker natural runoff. Deep channels may form as a result (Fig. 36). Avoiding the ensuing formation of badlands requires construction of a plurality of culverts for distribution of the runoff. It is worth noting that as regards quality, highway runoff differs from that of natural surfaces, since it contains materials emitted by the traffic (oil, grease, dirt, litter) and from erosion and dissolution of road construction materials (Sylvester and De Waller, 1972).

The "waterfall effect"

Most of the erosion caused to road structures results from the overflow at the borderline between the paved and unpaved surfaces. The absence of absorbing factors enhances the erosive potential, and after some time the eroded surface is lowered and a gap between the paved and unpaved surfaces is created. To the initial energy of the runoff flowing on the paved surface an additional energy is added by the force of gravity, because of the growing vertical difference between the paved and unpaved surfaces. Turbulence created by the falling water on the eroded surface exerts a "waterfall effect," leading to erosion of the road fill *beneath* the paved surface. If events are permitted to continue unchecked, the waterfall effect can cut off the whole width of the paved surface.

FIG. 36. Badlands created as a result of unsatisfactory construction of the highway Arad-Be'er Sheva, northern Negev, Israel (1972)

This waterfall effect can also demolish ditches, the gradient of which was supposed to be smoothed by small concrete platforms. An instructive case of this erosion, resulting from ignorance of basic geomorphological laws, happened along the highway mentioned in Figure 36. Small concrete plates (150 × 70 × 10 cm) were laid every 20–30 meters along the bottom of the 150-cm wide ditches near the road, apparently to smooth the 5–7% gradient. The smaller friction of those plates as compared to their surroundings increased water velocities, augmenting erosion around the plates by the waterfall effect, eventually undermining the plate. Today, those platforms are lying free, above the deeply cut ditch (Fig. 37), and what was intended as an improvement has had the opposite effect.

Siltation of channels

Heavy erosion in one part of a drainage system results, obviously, in silting of the other parts. Precipitated coarse material will impede flow in channels and culverts, causing overflow and inundations on and around the roads. This problem is connected with the problem of the size and type of bridges, discussed below (p. 84).

Problems of road construction in the permafrost zone

Engineering activities, be it road construction, building, airfields, pipelines (p. 90), etc., in the permafrost zone — northern Canada, Alaska, Siberia, northeast

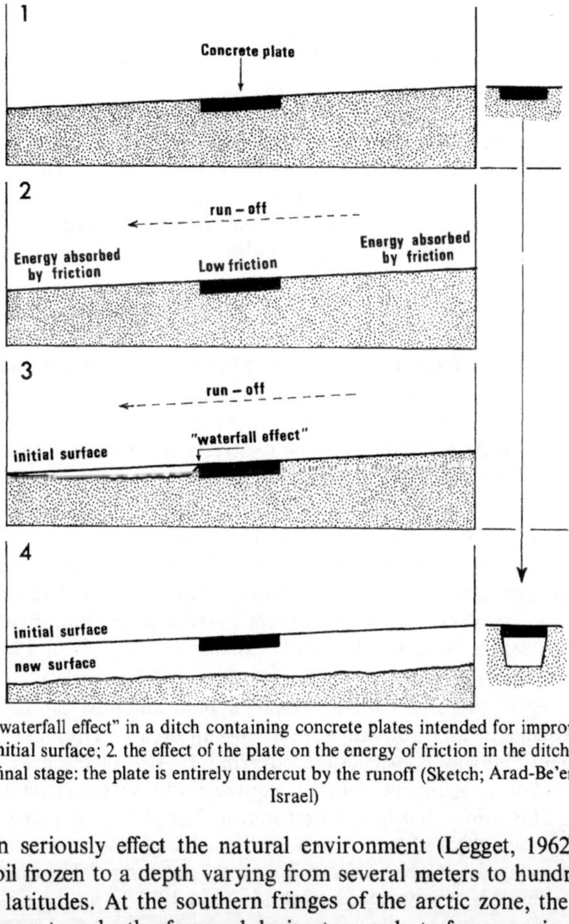

FIG. 37. The "waterfall effect" in a ditch containing concrete plates intended for improvement of stream regulation. 1. Initial surface; 2. the effect of the plate on the energy of friction in the ditch; 3. the "waterfall effect"; 4. the final stage: the plate is entirely undercut by the runoff (Sketch; Arad-Be'er Sheva highway, Israel)

Russia—can seriously effect the natural environment (Legget, 1962). Permafrost denotes a soil frozen to a depth varying from several meters to hundreds of meters at northern latitudes. At the southern fringes of the arctic zone, the top soil may thaw in summer to a depth of several decimeters, only to freeze again during winter. The thawing and freezing phenomenon is one of the difficult problems facing the arctic engineer. A permanent state of thaw, such as may result from constructions (roads, pipelines, and others), will have far-reaching consequences on the environmental forces (Demek, 1969).

Frequently, the frozen soil is protected by a cover of mosses, grasses, and sedges, underlain by a compressible mixture of partly disintegrated and decomposed organic materials, several meters thick and known in North America as "muskeg." Most of that layer is formed by organic debris of undrained or poorly drained depressions (Becker, 1972). Because of its fibrous character this material has an insulating effect, protecting frozen material from melting for all but a few decimeters beneath the plant cover.

When construction demands clearing the vegetation, soil temperatures are affected, and thawing penetrates deeper than before and may reach 3–4 meters (Becker, 1972). Once thawed, the soil material will not return to its initial state when refreezing, and an excess of free moisture results, changing the thermal and mechanical environmental properties. Subsidence will accompany extensive construction, demanding expensive counter measures. To prevent this undesirable phenomenon, the muskeg layer should be disturbed as little as possible (Legget, 1962). This approach has been employed along the Anchorage-Fairbanks highway (Alaska), where some 110 km ran through permafrost areas. The rate and extent of subsidence have been low enough to maintain the highway in good condition.

Erosion by Highways

There exists a basic difference between the phenomena of erosion accompanying each particular use of the land—agricultural, urban, or for transportation. Whereas agriculture-induced erosion, unless properly controlled, is a continuing phenomenon, construction sites erode mainly during the relatively short work period between the clearing of natural vegetation and the stabilization of the new surface (Guy and Ferguson, 1970).

It should, however, be borne in mind, that the road, once constructed, imposes permanent changes on slopes, hills, and valleys. The minimum gradients of modern highways only increase the amounts involved. Sherlock (1922) estimates that a two-lane highway involves over 300 m^3 of removed earth per km of road (in Britain). Assuming this average to hold, the roads existing in the U.S. in 1972, for example, would have necessitated transportation of over 7,000 million m^3 (equivalent to a cube having a side of 2 km). Removal of such amounts causes considerable erosion and sediment, as is illustrated in a 1.8-km^2 area in Virginia (U.S.) (Vice et al., 1969), where highway construction, covering 1–10% of the area at any time, contributed 85% of the sediment recorded, equivalent to 200 times the sediment from grassland and 2,000 times that from forest (on the basis of equal areas). An estimate (Collier, 1972) puts the sediment from a divided highlane in the U.S. at 1,200 tons/km^2, involving a surface of from 2.6 to 8.7 hectares. This would imply the erosional contribution of new roads in the U.S. in 1972 to have been 12.3 million tons of sediment.

The construction of new roads and highways is an overall phenomenon in modern society. New areas are being cultivated and the economic development of backward countries depends on reasonable transportation. Large-scale projects include the Trans-Amazon Highway, covering 5,300 km through Brazil from the Peruvian border to the Atlantic, crossing equatorial and tropical forests, which is a considerable potential cause for erosion unless properly managed.

Planners, engineers, and decision-makers should realize the effects of roads on geomorphological processes, especially during the construction period. The means required for preventing unnecessary soil erosion by road construction are basic and quite simple (Bennett, 1939):

1. reducing road side slopes, flattening of ditches;

2. planting grass on the road sides;
3. elimination of sharp angles over the transition from road to ditch;
4. reduction of the period of road construction;
5. establishing sufficient culverts, of appropriate size, to transport the discharge of a flood equal to the maximum measured over a period of at least 10 years;
6. reducing concrete-built ditches, and employing instead vegetation cover;
7. protecting road cuts by vegetation.

RAILWAYS

Railway tracks have been laid from the first quarter of the 19th century. The great expansion of the railway net took place in the third quarter of that century, with maximum annual new construction in the first quarter of the present one. Although some important new lines have been built lately (Tanzam, East Africa, 1,856 km; Baikal-Amur, Siberia, 3,200 km, under construction), there is a trend of decreased new construction with less viable tracks giving way to road traffic.

The main difference between road and railway, from the point of view of earth moving, lies in the low maximum slopes which a rail track may have. Larger-scale cutting than for a road of the same width is therefore the rule. In practice, the excavated material serves the rail track for making embankments, the amounts available and needed being about equal (Sherlock, 1922). In difficult areas, tunneling is of course employed. The amounts of earth which must be removed for building 1 km of single-line track have been estimated at some 70,000 m^3 (Sherlock, 1922). With about 1 million km of existing railway tracks (*Encyclopaedia Britannica*, 1974), it is seen that very considerable amounts are involved.

The erosive effect of the tracks, once laid, is restricted, as they are covered by loose gravel which well absorbs the mechanical effects of the rainfall and does not generate a concentrate runoff, provided that the number of bridges and culverts suffices. The restrictive character of this effect does not hold for the cutting, where erosion-controlling methods have not been used. The effect of accelerated erosion in these cases is disastrous, especially in semiarid lands, as has been illustrated by Aghassy (1973). The case study is a loess area in the northern Negev (Israel). A railroad system built prior to World War I by the Ottoman authorities connects Be'er Sheva with Gaza, crossing the wadi of Besor. After the war, the system has been dismantled, but remarkable erosional scars, developed into a true badlands landscape, are left across its path, especially over wadi crossings and cuts across hillsides.

Sliding on slopes from rail-track construction occurs on a greater scale than from those of road building. As a road copes with a steep slope through curves and serpentines, any local cut will have a limited bare area, even if the total area of all cuttings may be considerable. The railway track with a lower slope needs cuts of a greater exposed area, even if their number (per km) may be smaller. As erosion is a function, among others, of the length of the exposed slope (p. 40), its probability is greater on the railway tracks than along roads when equal areas are involved. The ferocity of erosion along the 170-km-long Assam-Bengal railway in India caused in 1915 damage necessitating repairs for two years, during which traffic was at a standstill (Legget, 1962). Another notorious example is that of the Folkestone-Dover rail track, which was extensively damaged in 1915 by a landslide caused by combined wave action from below and liberated water that was earlier contained from the side. Controlling those slides required protection of the lower slopes combined with good drainage systems, especially for marl, chalk, and clay formations, the angle of stability of which changes strongly with contained humidity (Sherlock, 1972). Concrete or stone/brick walls, even when drainage openings have been built in, are unsatisfactory. For really reliable service, pebble-filled gabions of steel wire are unsurpassed, assuring continuous and complete drainage.

BRIDGES

Bridges and Scouring

A construction of piers or pillars in a stream bed will decrease the cross-sectional area of the stream (Legget, 1962). As a result of this, especially if the volume of the pillars is relatively great with regard to the hydraulic cross section of the stream, the obstacle to the stream flow will induce turbulence, which is a major impetus to scouring. Moreover, for a considerable construction, stream velocities will increase at this point, and the water level may rise. Flow turbulence may give rise to eddies, which, combined with the increased velocity, will disturb the equilibrium of the bottom material, causing scouring of the waterway and damage to the bridge construction. Although known and preventable, the problem still takes far too great a toll of property and, occasionally, of lives.

Bridges and Sedimentation

Where the hydraulic cross section of a river immediately prior to a bridge is narrow, such a bridge may function as a dam, flooding the upstream area and causing there a deposit of silt. This phenomenon will usually happen when the discharge of the river surpasses the average on which construction of the bridge was based. A conspicuous example—and many exist!—is that of a bridge built in the 13th century by the Mameluke Sultan Baibars, just north of the town of Lod (Lydda), Israel. In 1954, Guy excavated the Roman base, to find no less than seven meters of silt, revealing the ancient bridge's function as a sediment trap, which undoubtedly caused inundations, to overcome which the 13th-century upper bridge must have been constructed. On the other hand, the previously mentioned "waterfall effect" caused considerable erosion on the downstream side of the bridge, amounting in one winter storm to 3,000 tons of clayey soil (Fig. 38).

Bridges and Bank Erosion

Where a bridged stream is not only too narrow to carry all of the volume of a flood, but its flanks near the bridge are also unprotected by an erosion-resistant construc-

FIG. 38. Bridge as an obstacle to natural flood. Wadi Ayalon, near Lydda, Israel, 23.2.1938 (photo: F. H. Taylor)

tion, heavy erosion of the banks at that location can occur. Figure 39 shows this type of bank erosion in a Mediterranean-type episodal river near Jerusalem. Heavy winter rains (1963) filled the bed, which became partly obstructed by floating trunks (right side of the figure). The stream attacked the other flank of the bridge (left side of the figure), where the protective wall was insufficient; serious erosion opened a meter-wide gap, leaving the bridge as an elevated construction and rendering it quite useless.

Bridges in Arid Regions

Bridges and culverts of unsufficient capacity to carry the floodwater in arid and semiarid regions are dangerous agents of erosion, devastating the highways they were built to protect. The problem is an economic one; building to the maximal probable discharge is expensive, and any smaller construction will be unnecessary during dry years and then fail in one single flood, when the too small passages will be blocked and the pavement damaged by the "waterfall effect." Under these conditions, the best solution consists of the so-called "Irish bridge," where the paved surface lies on the bottom of the river bed, leaving no vertical openings between the paved surface and the stream bed. Additional width, as well as paving with inserted pebbles, ensure perviousness to water, and an economically acceptable balance with nature is obtained. During the flood, no communication is of course possible, but little damage is caused to the pavement of the highway. Once the flood is over, communi-

FIG. 39. Man-made bank erosion, Nahal Soreq near Jerusalem, December 1963. The too small bridge, obstructed by vegetation, deflected the stream to the bank on the left side (of the picture), which was eroded away to render the bridge useless

cation can be resumed after cleaning the small volume of sediments left by the flood.

Erosion induced by inadequate bridges on a highway in an arid region, resulting in interruption of traffic and causing heavy damage to the highway and its surrounding, has been studied by Reichman (1976) on the Eilath-Ophira highway in south-eastern Sinai. The alignment of this highway was determined without sufficient data of the flood potential of the wadis crossed by this highway. Four flood events occurred during 1971 and 1972 (the highway was opened to traffic in 1970). Moreover, as on its crossing of the Wadi Wattir—a watershed of an area of more than 3,000 km^2—the highway has an insufficient number of culverts; it virtually dammed the wadi, resulting in overspilling of the highway and flooding of the nearby-situated village of Neviyot.

AIRFIELDS AND AIRPORTS

The number of airfields in the world is growing, as is the area occupied by each new one. London's Heathrow spreads out over 10 km^2, Paris' Charles de Gaulle over 30 km^2, and Montreal's new superairport has required expropriation of 356 km^2 (Danserau, 1971). Its construction involved moving some 500,000 m^3 of earth, with

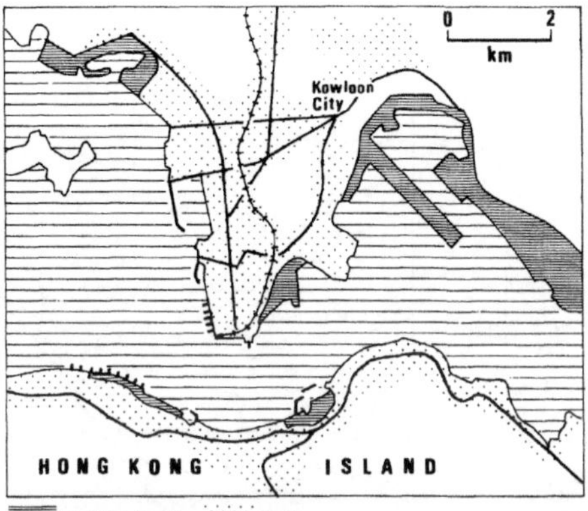

FIG. 40. Land reclamation from the ocean, Hong Kong (Chi-sen Liang, 1973)

the construction of corresponding drainage systems. The total number of airports, airfields, and airstrips is difficult to estimate; in the U.S. alone they surpass 10,000; many, however, consist of unpaved runways, and hence are of little influence on runoff intensity or direction. It is the immense metaled surfaces of the main airports which contribute concentrated runoff streams, with the consequences we have seen earlier. We know of no records of damage by erosion or sedimentation to the main airports; apparently the problems of drainage have been understood and treated well.

The need for ever greater level surfaces, distant enough from metropolitan centers to eliminate environmental effects but near enough to ensure rapid transit, has encouraged untraditional solutions, among which that of creating "terra firma" from the sea bottom. Hong Kong, with a limited and mountainous area, had its airport built in the sea on debris from building and land-improvement schemes (see p. 116 and Fig. 40). Many similar examples are known. The shallow waters of the shelf seem to be an easy target for such land reclamation. A British scheme off Foulness Islands, East Anglia, which would claim 36 km^2 from the North Sea (Fig. 41, p. 88), has not yet materialized (U.K. Dept. of Environment, 1972). It seems, however, that the need for additional, larger airports will continue to take up land in the vicinity of metropolitan centers in a growing degree.

WATERWAYS

Inland Waterways

The art of building canals for transportation is probably as old as that of digging for irrigation. Klimm (1956) assumes it to be one of Man's first large-scale interventions in his environment. The oldest canal discovered, in Mesopotamia, dates

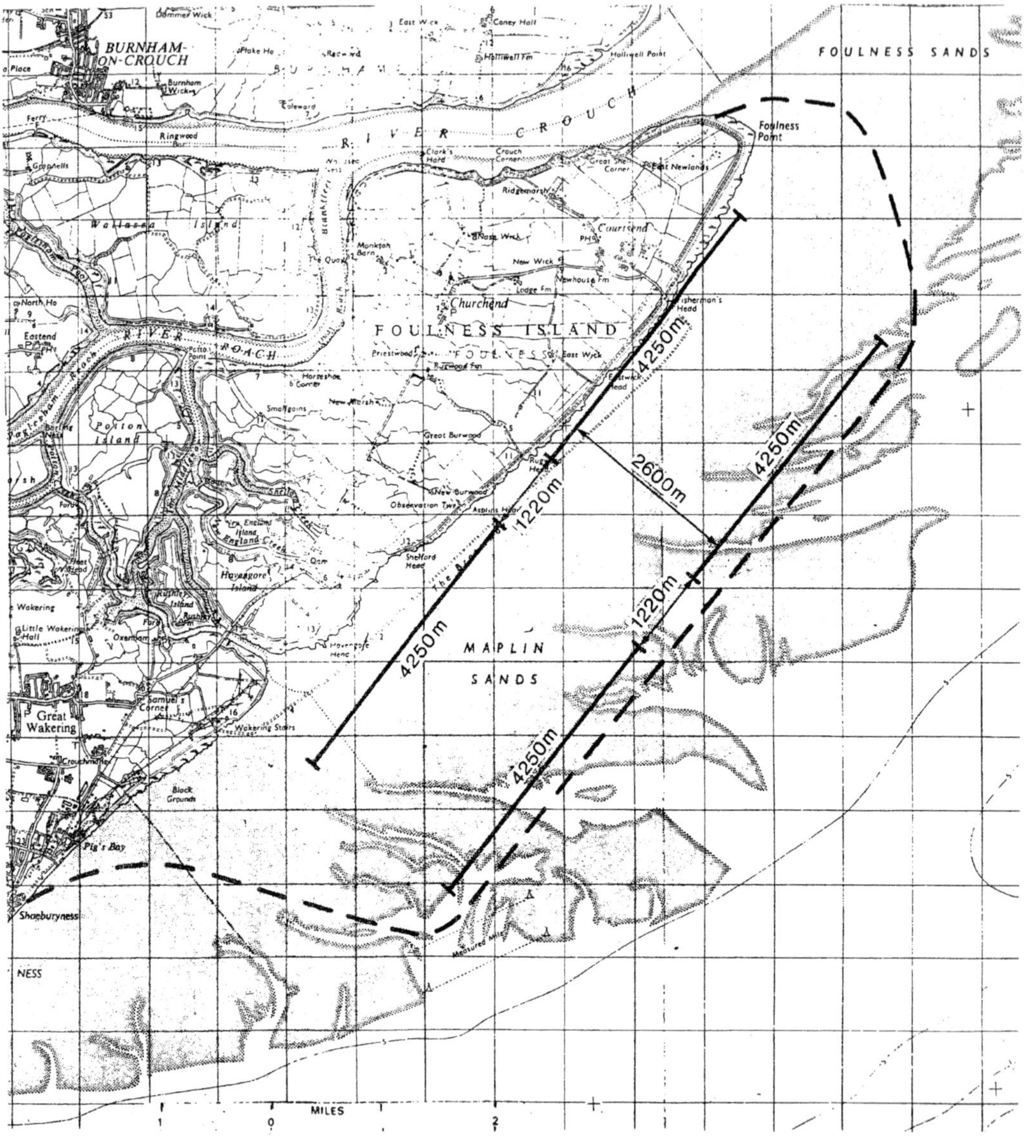

FIG. 41. The airport proposed for the Maplin Sands, East Anglia (U.K. Dept. of Environment, 1972)

from the year 5000 B.C. The 14th century B.C. witnessed a canal connecting the Nile and the Red Sea (Legget, 1962), a true forerunner of the modern Suez Canal. That century also saw a 135-km long canal in Kwansi, China, with others following. The important European canals date from the end of the 17th century A.D. and onwards. Their construction has involved hundreds of million of m^3 of earth (Sherlock, 1922; Fischer, 1915; Klimm, 1956).

Erosion of Canal Banks

Before the advent of steam and diesel engines, vessels were propelled by human or equine power, with corresponding low velocities and little damage to canal banks. Increased speeds have changed the picture. Experiments by de Salis (1904) showed that "no waves of an injurious character appeared at a speed of 3 miles (4.8 km) an hour," and that "between 3 and 3.5 miles (5.6 km) a breaking wave appeared occasionally in curves and shallows." The maximum velocities allowed today, such as 8 miles per hour (12.9 km/hour) in the U.K., certainly create an "injurious wave" by the ever-growing number of vessels of all kinds using the waterways. Erosion of the banks, especially in the curves (or in the meanders of the rivers) is evident, requiring constant maintenance of the banks as well as of the navigable rivers to prevent erosion which might attain disastrous proportions.

Canals through Isthmuses

Canals through isthmuses, shortening sea routes, have been constructed in the last 120 years. Three important such canals are those of Corinth, of Suez, and of Panama.

The Canal of Corinth—opened to navigation in 1893—is a relatively small canal, 6.4 km long, 23 m wide, and 9 m deep. It passes through tough limestone strata, and hence no erosion or slides of importance have occurred (Fig. 42). The Suez Canal passes through quite a different environment, traversing as it does 162 km of sand dunes and saline marshes. Its average width is 60 m, and its depth 12 m, indicating some 110 million m^3 of removed material; in the late 1970s, a new program of deepening and widening has been started. Since its opening 120 years ago, two geomorphological factors have joined in refilling the canal by sand: transport by wind and—to a greater degree—sliding from the banks as a result of waves generated by passing ships. Constant dredging is required to keep the canal in shape, as has been shown during periods of forced nonmaintenance during the Middle East wars.

Of the three canals, the greatest geomorphological impact has been caused by the Panama Canal. It was built from 1882 to 1914, with a total length of 64 km. The terrain traversed is difficult, and required cutting through a relief over 100 m high. The steep slopes of the unstable material to be traversed, combined with a wet, hot tropical environment, gave rise to landslides of such dimensions as to cause interruption of the work. So, for example, one single slide of a volume of 382,000 m^3 has been reported to happen overnight. Final shaping required removal of 675

FIG. 42. The shipping canal crossing the Isthmus of Corinth

million m^3 of earth, nearly 4 times the total annual sediment load of the Mississippi River (Collier, 1972). It should be remembered that this huge transport occurred on an extremely limited area.

PIPELINES

The number and extent of pipelines crossing both the developed and the developing countries increase from day to day. Compared with other surface transport modes, one would tend to consider the disturbances caused by pipelines to be temporary and to disappear once they have been buried. In view of the tendency to lay the pipeline over as short a track as possible, disregarding topographical features, the impact is bound to be heavier than that of roads of rail tracks. Furthermore, the impact will be limited only if the original equilibrium is restored or if proper measures are taken. In regions where the land is of low value, environmental considerations may depend on the presence or absence of a strong public opinion.

Generally speaking there are two kinds of environments where erosive hazards are serious. One is the arid region, where no natural vegetation will grow to prevent wind and gully erosion. The other is the arctic region, where permafrost conditions (pp. 81, 82) have a lasting influence. Construction under arctic conditions was studied

.vhen planning the Alaska pipeline (Palmer, 1972). The effect of transporting a hot fluid (14°C is the preferred temperature) will be to thaw permanently the surface carrying the pipeline. Under Alaskan conditions, a thaw to 4–6 meters, attained in 2 years (and continuing to 9 meters after 9 years) would cause such subsidence as to break the pipes. It was therefore decided to construct the pipeline above the ground, resting on pillars standing 30 meters apart, anchored to such depth as to be independent of any seasonal thawing. In this way, the natural steady state—the permafrost—will be unaffected by the operation of the pipeline.

One additional difference between the passage in landscape of a pipeline, compared to the other means of surface transports, is its absolute independence of topographical features. A pipeline is traced in the shortest possible line, crossing valleys and mountains, when there is a technological possibility to cross them and when the subsequent maintenance of the transport of the fluids (i.e. pumping) is possible from the economical point of view. Therefore, the erosive effect of a scar left after the laying of a pipeline is more intensive than the construction of a road or a railway track, the latter relying on a certain control of slope and relief steepness, avoiding steep gradients; the pipeline's scars, cutting the relief by straight lines, are scrupulous interventions into the Earth's surface and their contribution to soil erosion should be feared.

6

Riverbed Management

INTRODUCTION

Riverbed and floodplain management aims either at prevention of floods or at rehabilitation after damage. The scale of operations—local, national, or wider—will depend on circumstances. The effects on downstream localities should not be discounted when considering any improvement. Events in Kuala Lumpur clearly bear this out. The Sungei Gombak River was regulated throughout the township, thereby preventing floods or erosion. Downstream, however, the untreated banks were subject to heavy bank erosion, greater than that prior to intervention upstream, because of the increased velocity at the point of leaving the regulated banks (Douglas, 1976). The lesson from this case, which certainly is not extraordinary, is that geomorphological control of a river should take into account the whole watershed and its hydrographical system.

Regulation and management of a river can be carried out by two approaches (Strahler and Strahler, 1973; Keller, 1976); either by the treatment of the slopes, which includes afforestation to reduce runoff and to increase infiltration and terracing of fields, or by direct protection of the flood plain, by widening, deepening, and straightening of the stream bed, cleaning of it by dredging, protecting it by dike building, and decreasing flow rates by temporarily storing water in reservoirs for controlled release.

CHANGING RIVERS BY REGULATION

One may safely assume that there are no "natural" water courses left in the developed countries, irrespective of the use of adjacent land. Although it is mainly the undertakings of the last few decades that are known, through recent documentation or larger effects, sporadic observations of Man's intervention in river systems have been made throughout the history of mankind.

Dikes and Levees

One of the most ancient human interventions in the course of a river is that of the Nile (Holz, 1969). Radiocarbon dating indicates that agricultural settlement in the Nile Delta began about 7,000 years ago, over which period we know that the pattern of the delta changed considerably. Another example of a continuous engineering effort to harness the water—and the silt—of a river is that of the Yellow River in China. On flowing along unstable loess-covered banks, rains in higher reaches cause floods loaded with silt in the lower, densely populated agricultural and urban areas (Eliassen, 1936). The river transports annually some 1,500 million tons of silt, a quantity equal to the load of all the U.S. rivers taken together. The heavy load leaves little energy for erosion by the stream. The gradual deposition of the silt requires construction of dikes to contain the water. At increased seasonal flows even those dikes are insufficient, and areas outside are flooded to a height of 3–5 meters.

Another river having extensive dikes is the Mississippi, the total length of which exceeds 6,000 km. Its flow at Vicksburg (some 450 km from its mouth) averages 14,200 m^3/sec, but this may—during a flood—reach 56,600 m^3/sec, an increase of 300%. The first attempts to harness the Mississippi, during the 19th century, included partially raising the natural levees by adding artificial embankments. The local protection was, however, obtained at the expense of worse floods caused by higher water levels elsewhere. In other words, no local additions to the height of the levees were effective unless levees were raised simultaneously over considerable lengths (Legget, 1962). At first, levee height was 3.35 m; in 1931, the average height had gone up to 4.5 m, and that over a length of 3,700 km.

Mere construction of dikes along rivers may have disastrous results if a sudden flood rises above their level. A new discharge and a totally new river course may result. This is what happened at the Yellow River in 1892, when a flood broke the banks and created a new mouth 450 km to the south of the former one (Legget, 1962). A similar effect occurred in The Netherlands in 1421 (Wagret, 1968). In 1270, the Old Meuse River had been dammed in, its course being deflected towards the Rhine, with the object of creating a new polder. An unusual discharge broke the dams and caused, in Wagret's opinion, the catastrophe known as Sint Elisabetsvloed, in which, in one single night, 42,000 hectares of fertile land were submerged, 72 villages were completely destroyed, and 100,000 people drowned (other sources mention less casualties and damage).

In addition to their height and length, the resistance of levees to erosion is of importance. Suitable control of the slopes where danger is greatest is now considered essential. Protecting the earthen dikes of a stream from erosion is even more problematical in arid regions, where, because of the absence of long-term records, the magnitude of floods and their probable frequency are practically unknown. Deciding on the dimensions of a dike is therefore extremely haphazard. On the other hand, the dikes, by their sheer volume, may create a false sense of security and encourage the construction of settlements and the development of agricultural activities prone to great dangers when such dikes collapse and release huge quantities of stored water. In one such recent, recorded case in eastern Sinai a pierced dike collapsed

70 minutes after the flood began, releasing a water front 200 m wide and 5 m high, which submerged an 8-hectare area for 7 hours, destroying water pipelines and a metaled highway (*Ha'aretz* daily newspaper, 26 November 1972).

Shortening Rivers by Channels

The shortening of rivers by digging channels has been carried out on a wide scale. In the U.S. it has been a Federal Government activity for well over 100 years (Gilette, 1972), involving over 15,000 km and effecting the drainage of over 4.5 million hectares. Several investigators have examined the geomorphology of the streams, as affected by the channeling activities (Daniels, 1960; Ruhe, 1971; Slaymaker and Gilbert, 1972; and others).

A well-documented case is that of the Willow River, Iowa (Daniels, 1960; Ruhe, 1971). In its flow over 42 km through a valley, its gradient varied considerably, from 0.8 to 2.4 m/km, as a result of local overloading by sediment through the tributaries.

Regulation of the river included shortening, widening, deepening, and decrease of variation in the gradient (Daniels, 1960). Serious erosion in the upper parts resulted with a corresponding sedimentation below, requiring additional subsequent dredging. Erosion worked headwards, moving the knickpoints backwards up to 30 and more m/day, especially after periods of heavy rains (Fig. 43).

These drastic changes in the river are undoubtedly the result of the change of a single factor without consideration of others. Preserving the original gradients combined with artificial meanders would have retained the original steady state.

Such considerations directed the improvements in the Otoe Bend of the Missouri River (Ruhe, 1971), where the original gradient was retained, smooth artificial meanders were constructed, additional land for agricultural purposes was created, and navigation was improved (Fig. 44). No significant sedimentation or erosion below or above the Bend have been observed.

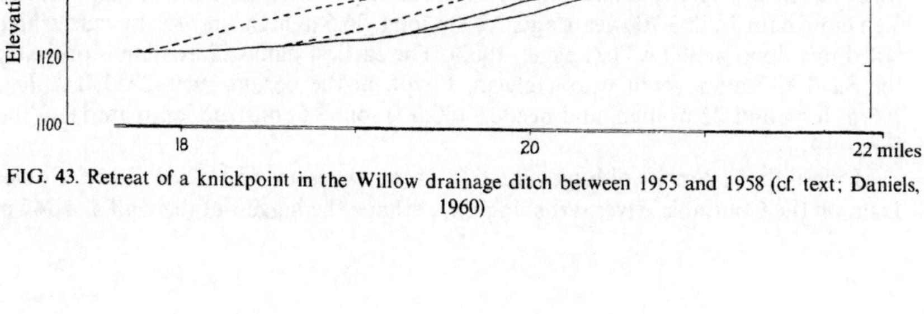

FIG. 43. Retreat of a knickpoint in the Willow drainage ditch between 1955 and 1958 (cf. text; Daniels, 1960)

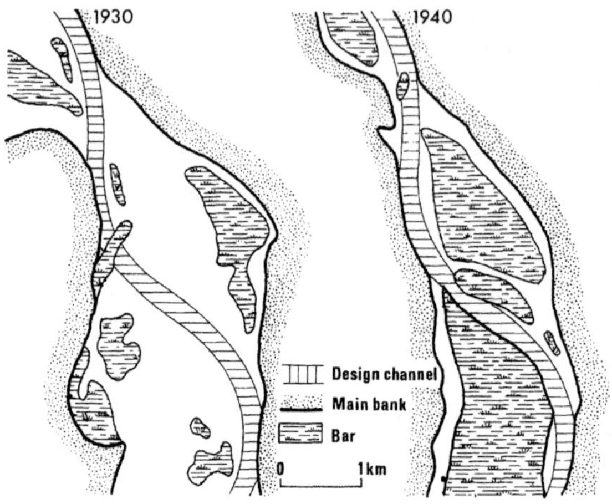

FIG. 44. Regulation of the Otoe Bend area, Missouri River, between 1930 and 1940 (Ruhe, 1971)

DAMS

Introduction

The building of dams is undoubtedly one of the most ancient human engineering activities, and as such, closely related to the regulation of irrigation and navigation (Legget, 1962). The oldest-known dams were built over the Tigris, the Euphrates, the Nile, and the Yellow River. The recent increased needs for electric power, as well as for control of erosion, have added high and wide dams, first in the Western world, and later as well in the developing countries of Asia, Africa, and South America. Today, the damming of rivers for irrigation, water storage, erosion control, hydro-electrical purposes, or a combination of these factors is a general practice. The percentage of total runoff thus regulated is 21 in Africa, 15–20 in the Western world and less elsewhere (Beaumont, 1978). The first masonry dam of which there is a fair record was built over 6,000 years ago over the Nile, south of Memphis (Legget, 1962), with a length of 457 m and a height of at least 15 m. It served for 4,500 years before yielding to neglect. Another known dam, constructed in prehistoric times at Marouk on the Tigris, lasted till the 13th century A.D. Less known, perhaps, is a Yemenite dam, built 3,700 years ago, 3.2 km long, 36.5 m high, and 152 m wide, which failed in a flood in 300 A.D. (Legget, 1962). The earliest dam still existing is probably the Saad el Kafara, built near Helwan, Egypt, in the period 2950–2750 B.C. It is 106 m long and 22 m high, and needed 100,000 tons of construction material (Vita-Finzi, 1974).

Modern dams consist of huge concrete, stone, or earth walls. The Grand Coulee Dam on the Columbia River, Washington, perhaps the biggest of them all, is 1,247 m long, 167 m high, and weighs 20 million tons. Dams create artificial lakes, one of the largest being Lake Volta in West Africa, with a surface of 8,482 km^2. Another giant, the Kuybishev Dam Lake in the U.S.S.R., occupies 6,500 km^2 and is one of 50 artificial reservoirs created on the Volga-Don-Dnieper inland water system (Makhaveyev, 1972). The volume of water retained by these reservoirs may be tremendous, such as the Bratsk reservoir on the Angara River in Siberia, containing 170 milliard m^3 and the Kariba Dam in South Africa containing 160 milliard m^3. Where a lake's outlet is dammed, its water level will rise, bringing with it additional hydrological effects. Owen Falls Dam raised the level of Lake Victoria by one meter, adding 204 milliard m^3 to the storage. Where rivers flowing into lakes are dammed, and their water is used, the area of the lake will decrease, as was the case with Lake Aral, Siberia, and the Caspian Sea, decreasing by 38,000 and 25,000 km^2, respectively (Vita-Finzi, 1974).

Opposition to unlimited dam construction, which disregards environmental considerations, has been voiced by Kollmorgen (1953), who speaks of "floodplain cannibalism" where a multiplicity of dams stepwise cover the major part of valuable alluvial valleys. The Grand River, Missouri scheme, given as an example of such a case, ultimately required inundation of 61,110 hectares to protect 85,800 hectares elsewhere, with the additional costs of displacing 5 towns and hundreds of farms and depriving the floodplain of the fine alluvial sediment which was held back by the dams.

It is clear that a more detailed geomorphological examination of the effects of dams is required in order to comprehend the problems accompanying their construction.

Dams as Factors of Accelerated Erosion

Dams consist, by definition, of a human intervention in a river, smoothing the gradient of the river upstream, and making it steeper downstream (Fig. 45). The increased hydrological cross section of the artificial lake, resulting in decreased flow velocities, stimulates sedimentation. The water passing over (or through) the dam, containing a smaller sediment load, erodes the new river bed, which phenomenon is enhanced by the "waterfall effect." Erosion will continue as far and as long as the flow has not acquired a sediment load to reattain an equilibrium with the river bed.

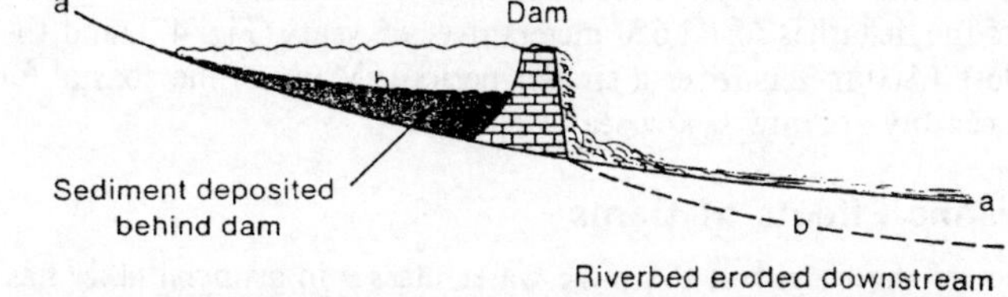

FIG. 45. Sedimentation and erosion accompanying a dam. a. River-bed prior to construction; b. same after dam has been operating for some time (Collier, 1972)

The long-term effects of a dam can thus be summarized distinctly for the three river parts involved: *upstream* from the edge of the lake, where regressive aggradation gradually spreads up the river, resulting in a new valley floor; *inside* the reservoir itself, where waves, currents, seasonal changes in the water level, and gravity processes combine to produce a distinctive shoreline relief, such as cliffs, lagoons, and bays; *below* the dam, where the water carries little or no sediment and is capable of eroding the riverbed downstream. The scouring of the river channel causes formation of a terrace (parallel to the flow direction), the relative height of which decreases from the vicinity of the dam downstream, disappearing where the scouring activity ends.

Another, apparently contradictory effect below the dam consists of a progressive reduction in channel dimensions. It is explained (Richards and Woods, 1976) by the "absorption" of the peak floods in the reservoir upstream. The actual situation below the Burrator reservoir near Dartmoor, United Kingdom, for example, is an illustration of this state.

Dams' Influence on the Distribution of Sediment

Although it is obvious that a climatic and environmental study of a river and its watershed should precede the construction of a dam, this has at times been neglected, with the result that accumulating sediment, and especially its coarser components, has been filling up the reservoirs. In one, admittedly extreme, case, 95.6% of the reservoir volume were filled—and therefore lost—in 13 years of operation only (New Lake, near Austin, U.S.). Ultimately, the sedimentation process will extend upstream into the river bed proper.

The described process of sediments accumulating behind a dam may be marked in semiarid regions where the silt content of the flowing watere is the highest in the world (Guilcher, 1965). A filled-up reservoir, when subjected to a sudden heavy rainfall, may cause an entire dam to collapse, as happened in 1968 at Ein Khudeirat, in northeast Sinai, after a mere 30 years of service (Fig. 46). Partial release of silt-rich water through an appropriate sluice might have prevented the mishap.

The sediment trapped behind a dam is, of course, deprived from the river in its lower course. This may have far-reaching results on the coastline at the mouth of the river, as has been illustrated at the Rosetta branch of the Nile River Delta. Until the Hellenistic period, settlements moved northward, as indicated by archaeological evidence. In the last 100 years, however, increasing construction of dams—culminating in the Aswan Dam, 1970—deprived the delta of its sediment load, whereas maritime currents have retained their transportation capacity. As a result, the Rosetta branch of the delta has lost 1,650 meters over 65 years (Fig. 47), and the Damietta branch lost 1,800 meters (over a similar period). Many of the coastal forts built in the 19th century are now below sea level.

Seismic Effects of Dams

The effect of the weight of the huge water masses in artificial lakes has been considered by many, who usually based themselves on the measurable physical quantities (weight, permeabilities, tectonics).

FIG. 46. Ein Khudeirat, northeast Sinai. Accumulated silt over a period of 30 years, combined with a heavy flood in 1968, caused the dam to collapse

Case studies on several dams (International Colloquium on Seismic Effects of Reservoir Impounding, 27–29 March, 1973, in *Eng. Geology*, **8**, pp. 1–212, 1974) lead to the conclusion that earthquakes might be triggered by artificially induced changes in fluid pressure in the crust of the Earth in the vicinity of large man-made reservoirs. On the other hand, there are thousands of reservoirs where seismic activity has not been noted. Therefore, the causes for a seismic event, induced by a reservoir, may be as follows (Judd, 1974):

a. The reservoir water load alone could be sufficient to initiate seismic forces;
b. The crustal stresses would have to be at a critical stage before the rather nominal load imposed by the reservoir would trigger seismic events;
c. The reservoir water permeating the underlying strata may increase the in situ pore pressure sufficiently to decrease the effective normal stresses and thus diminish shear strength along any local fault planes;
d. The reservoir water may cause an increase in hydrostatic pressure in the underlying crust and thus produce a stress redistribution of sufficient severity to initiate the seismic shock;
e. A combination of pore pressure and saturation may sufficiently reduce the strength and elastic moduli of the rock masses to facilitate release of crustal strains.

The collapse of the Malpasset Dam in southern France (1959), devastating the town of Fréjus on the Côte d'Azure, has been attributed to a fault in the base rock beneath the dam. Whatever the true cause may be, it is clear that our knowledge is

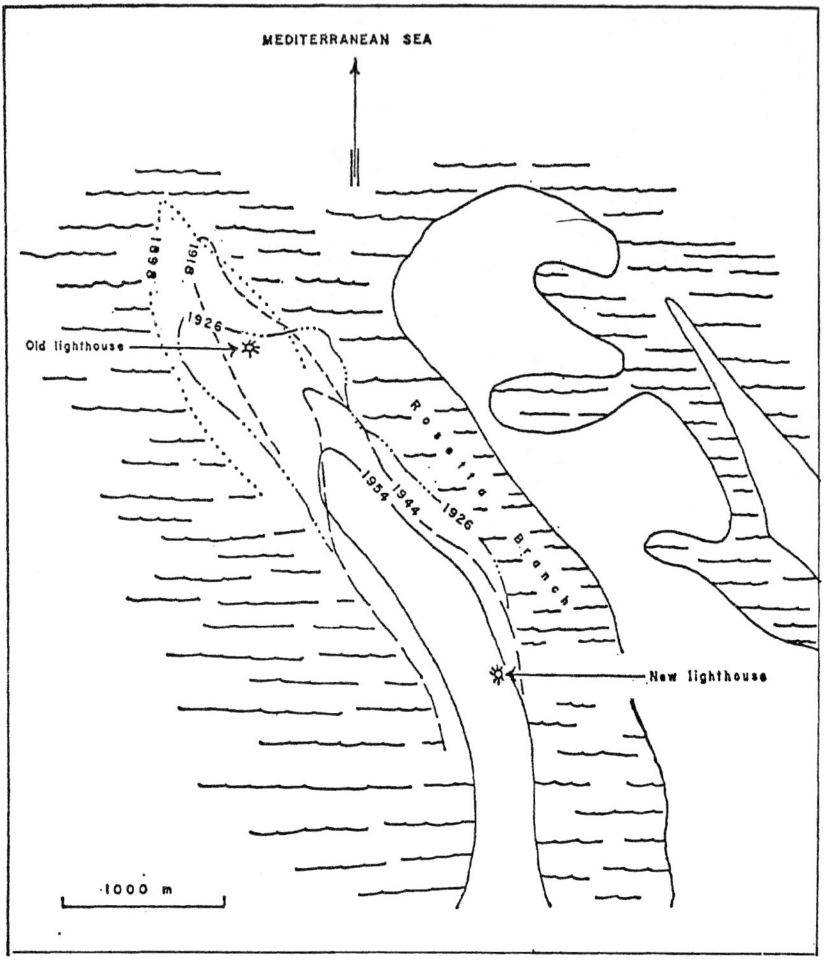

FIG. 47. Changes in the Rosetta mouth of the Nile Delta, caused by the establishment of dams on the river (Kassas, 1972)

insufficient to understand the implications of such a permanent redistribution of weights on the lithosphere. Legget (1962) has established that Lake Mead on the Hoover Dam, Colorado River, where some 38,000 million tons of water are concentrated, has caused a subsidence of land around the lake of up to 20 cm over 20 years. Furthermore, several small but observable earthquakes have been noted by an adjacent seismograph (Boulder City), apparently resulting from the same cause.

Other man-made earthquakes seem to have occurred in the Kariba Dam region, with an intensity of 4–6 (1967), and in the Sao Paulo State, Brazil in 1966, near the

Barrage de Foros. In both areas mentioned, no seismic shocks were ever recorded before the dams were built (Carder, 1968). Other cases which have been examined include Marathon Lake, completed in 1929, where the shocks were first felt in 1931 (Božovič, 1974); Pieva Di Cadore Dam in Italy, built in rigid carbonate rocks, where some larger shocks have been reported in 1964 (the dam was built in 1949); Granval Dam in central France, built in 1959 in a region known as to be seismically quiescent, where a violent shock with hypocenter below the storage was recorded in 1963, just after the filling of the reservoir, etc.

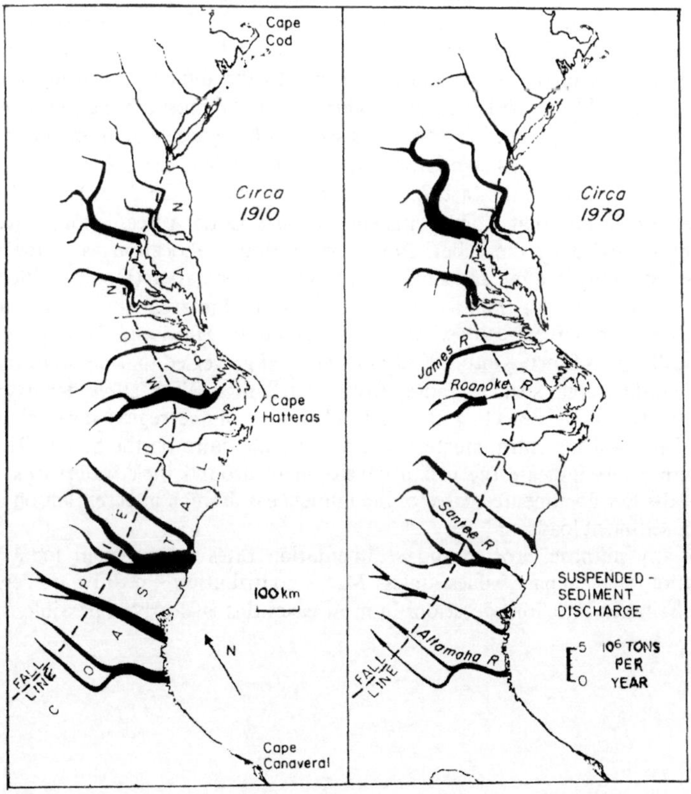

FIG. 48. Average annual loads of suspended material carried by U.S. rivers flowing to the Atlantic Ocean. Left — the load in 1910; right — the load in 1970 (Meade and Trimble, 1974)

CHANGES OF WATER COURSES AND SEDIMENTATION RATES

It is difficult to underestimate the degree of human intervention in the course of the rivers as they flow today. Louis (1960) states that there exist no untouched rivers in Western and Central Europe. In fact, rivers have been removed from the catchment of one ocean to that of another; witness the "Big Thompson Scheme" near Denver, Colorado (Legget, 1962) and the "Long Lake" project in northern Ontario, Canada.

Further intervention results, undoubtedly, from the considerable volumes abstracted to irrigation, calculated (see p. 49) to amount to some 1.6×10^{12} m^3 annually. The quantitative expression of this intervention and of its effects on hydrological and environmental factors is, in fact, one of the objects of the International Hydrological Program (Unesco, *Nat. Res.*, **11**, pp. 26–27). Where we try to evaluate the influence of Man's intervention in rivers and their watersheds on the sedimentation rates in the floodplains, we run into the difficulty of being unable to separate natural from artificial causes (Dury, 1974). In most areas of settlement and cultivation in floodplains, Man's actions seem to have accelerated, rather than triggered, the natural process. We know that the Rhône delta has grown (outwards) by some 11 m/year over the last 2,300 years, and that of the Danube by 3 m/year (Legget, 1962), and that Rome's ancient sea port, Ostia Attica, is long covered by sediments carried by the Tiber. But no quantitative allocation as to the various causes is possible in these and similar cases. One investigation (Douglas, 1967b) puts sediment yields of Australian rivers at 80–750 m^3/km^2/year, of which only 4–10 would be from natural causes. Another calculation (Mead and Trimble, 1974) on some U.S. rivers flowing into the Atlantic (Fig. 48) revealed that the southern rivers in the studied area, such as James River and Santee River, show decreased sediment loads, clearly reflecting the rural character of the region covered by their watersheds and the entrapment of sediment by the dams on the Santee River. The northern rivers indicate the urban development around their watersheds. This is one of the few documented cases of the influence of human intervention on changes in river sediment loads.

For any purpose, erosive and sedimentation rates as measured today are no indication of their past values, unless Man's contribution—positive or negative—could be taken into consideration. In most cases this is clearly impossible.

7

Shore Management

INTRODUCTION

When the ancients built the first shore constructions to facilitate arrival and departure of ships in river mouths, shore management may be considered to have started. There is, of course, much more to it, bearing relation to protection of the coast against erosion, reclamation of inundated land, marshes, and swamps, and the mining of building materials such as sand, shingles, and gravel. All these human activities constitute an intervention in a complex system influenced by oceanographical, climatic, and geomorphological factors. Sea shores are, geomorphologically, a most active zone as they are subject to considerable energy impact in the form of storms and eddies, concentrated over narrow strips and in short periods. As a result, the shores are continually changing, more so than any other region.

The power which artificial installations have to withstand is considerable. Vertical obstacles to waves will undergo action by the following power (Williams, 1960): on a calm day, waves of 30 cm height and of a 30-second period will develop 0.9 kW (1.2 HP) per linear (coastline) meter, whereas on a stormy day, waves of 7 m height, 300 m length and a period of 14 seconds will develop 170 kW/m (230 HP/m).*

The ability of waves to cause damage is well documented. Take, for example, the Port of Wick in northern Scotland (Legget, 1962). In 1872 a breakwater was destroyed and pieces of construction weighing up to 1,350 tons were carried by the waves. After repair, a storm in 1877 carried off blocks up to 2,600 tons! Of course, local measures to prevent similar events can be (and have been) taken. It should, however, be noted, that as with other interventions in natural processes, a change introduced in one region will have repercussions in other, adjacent ones.

* In William's (1960) formula, the developed power is directly proportional to gravity, water density, wave amplitude, and wave velocity.

PROTECTION OF COASTS AGAINST EROSION

The best protection of coasts against erosion consists of the natural one, by the accumulated shingles and sand (Sherlock, 1922). The natural comminution of the shingles, combined with their mining—with sand—as building materials, brings about their depletion, and causes a need for further, artificial protection.

Seawalls

One of the most universal means to protect coasts is by seawalls, which may also serve as basis for roads and promenades (Royal Commission, 1911). Various profiles of seawalls are in use. It is of cardinal importance, however, that these conform to the demands put by the waves in the particular region concerned, as otherwise the seawalls cause the receding waves to scoop out the sand at their base, the walls serving as "agents of their own destruction" (Royal Commission, cited above). Groynes (see below) may help in preventing the sand from eroding away. To prevent destruction through erosion, which attacks first where the resistance is lowest, seawalls should be founded on the firm bed of rocks, and not on the sand or shingles. Seawalls are also sometimes endangered by water which penetrated to their backside, starting a scooping-out process where it is least expected.

Groynes

A groyne consists of a framework of low broad wall, usually made from timber, boulders, or concrete, run out more or less perpendicularly to the shoreline, to check drifting of the beach. For greatest efficacy, it should extend from the seawall without any intervening gap, and begin well above the high-water mark (Royal Commission, 1911). The length of the groynes, their distance from one another, and their exact angle with the shoreline will vary from one locality to another. The *length* should be at best that needed to "cover" the region of the breakers and of the currents along the shore, and will usually amount to between 50 and 200 meters. The groynes' *height* should be at least 0.5 m above high-water level for sandy shores, and at least 2.0 m above it for shingly ones, the greater height for the latter resulting from their lesser capacity of absorbing the energy of the waves as opposed to the easier drifting of the sand. The *distance* between the groynes should not exceed their length. Where this distance amounted to 1.5–4 times their length, as in parts of New Jersey, the beach receded 0.6 m/year over 80 years (Legget, 1962). The *angle* with the shoreline will depend on the prevailing winds, with which the groynes preferably will be perpendicular, taking further also the currents along the shore into account.

In the matter of groynes, much has been learned from the seawall construction at Scarsborough (U.K.). The walls' apron permits the striking sea to rise, with waves reaching at times 50 m height (Royal Commission, 1911), causing heavy erosion on their retreat. Addition of groynes has completely reversed this process (Fig. 49), and additional sand layers have been formed.

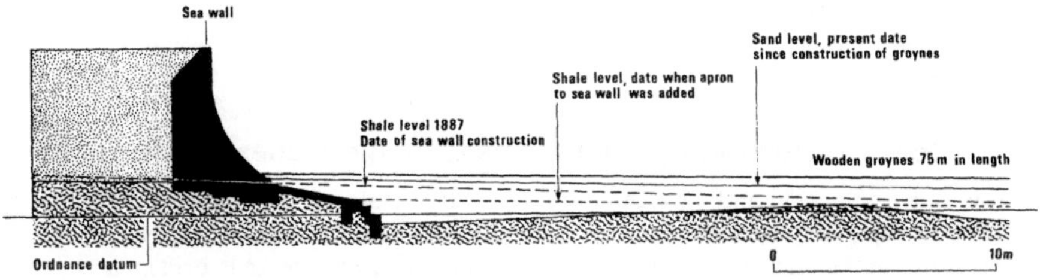

FIG. 49. Effects of seawall and groynes on the scouring and the filling along the seashore, Royal Albert Drive, Scarborough, U.K. Construction of the groynes caused the sand to settle (Royal Commission, 1911)

Over the last decades groynes have been employed to improve on beaches of value for the tourism industry, mainly through providing additional beach surface. The northwestern Mediterranean alone contains tens of such locations. In many such cases breakwaters *parallel* to the shore have been constructed. The unintentional result may be a *tombolo*, a sand or gravel bar connecting the breakwater with the mainland. One such case occurred on the shore of Netanya, an Israeli resort (Fig. 50). A breakwater to protect the beach against western breakers was constructed;

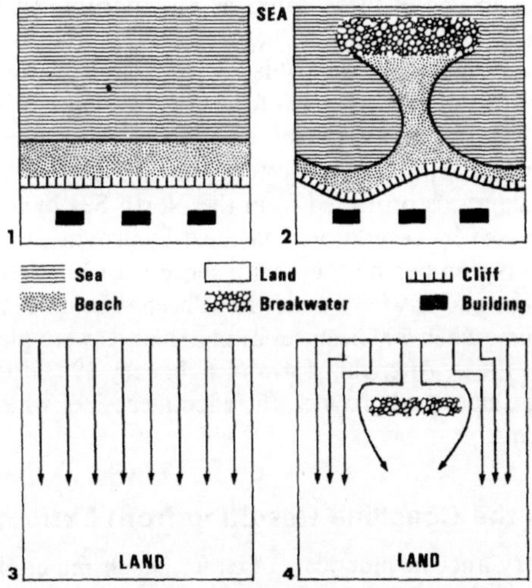

FIG. 50. The influence of a breakwater on the configuration of the coast of Netanya, Israel 1) (Left, top): Coast prior to any interference; 2) (right, top): after construction, with changed position of sand and coastline; 3) (left, bottom): forces affecting the coast before interference; 4) (right, bottom): forces operating after the construction

the changes in the pattern of the currents formed a tombolo and caused heavy erosion of the adjoining cliffs, seriously endangering a new hotel built on top (*Yediot Aharonot* daily, 16.6.1971).

Coast Protection by Improvement of the Dunes

The rule that each and any interference with a natural process will have far-reaching repercussions has been encountered often in the present work. Nowhere is this more valid than in the case of the seashore. The ever-present drifts, currents, and waves change daily as well as with the seasons, even if over extended periods some long-term equilibrium may appear. Man's tampering with this apparent instability has yielded results both unexpected and extensive, as, for example, on the eastern shores of the U.S. The Barrier Island is part of North Carolina's off-shore (barrier) islands. Its sands do not rest on any parent rock. To protect a proposed road from inundation by the sea, a dune system was erected by moving some 1 million m^3 sands; the system was stabilized with trees and shrubs (Dolan, 1973).

Thirty years after construction this part of the shore demonstrates the rule of the indirect results. The width of the beach here is 30 m or less, as compared with 125 to 200 m elsewhere. It is probably the steeper slope—similar to the effect of the seawall—which causes increased turbulence of the receding waves, resulting in increased erosion and a steeper beach profile (Godfrey and Godfrey, 1973).

Godfrey proposes to accept these banks as an environment made up of flexible factors, such as sand, waves, wind, and natural grasses, and to treat it accordingly. It may then form an ecosystem well able to withstand continuous natural stress, expected to maintain itself in the present form for the "foreseeable future."

There exist examples of dunes serving as effective coast protectors. One case is The Netherlands, where most of the western coast, from the Rhine-Meuse estuary up to the Frisian Islands, is protected from the North Sea by a strip of sand dunes. Covered and buttressed by vegetation, protected by groynes, these sand dunes have been protecting the region behind them with success for hundreds of years. Another case is the United Kingdom, where dunes have been safeguarding the Norfolk coast since the 17th century, when 800 hectares of marshlands were planted with maritime pine trees (Wagret, 1968). After the storm of February 1953, these dunes were buttressed by concrete slopes and groynes, the effectiveness of which was satisfactorily proved in later storms.

Changes in the Coastline Resulting from Extraction of Materials

It certainly is a very ancient practice to extract beach materials — shingles, sand, and gravel—for building and other activities. According to legend, glass manufacture originated along the sand-bearing shores of the eastern Mediterranean. Even shells are extracted and employed after crushing as a source of calcium for poultry (Newe Yam, Israel), or, when mixed with sand, as a base for road construction (Liberty Country, Texas; Legget, 1962). The extraction of materials is accompanied, however, by erosion of coastal areas, and the two phenomena seem to be correlated.

In some instances, definite proof of cause and effect is at hand. One such case is that of the region of Street Head, Devon, England, where in the period between 1897 and 1901, 650,000 tons of shingle were mined over a stretch of 8 km. In 1901 the sea started encroaching on the land, and in three years, great damage to houses, seawalls, and roads was caused, necessitating a complete stoppage of all mining activity (Sherlock, 1922).

A similar case of accelerated shore recession following upon mining occurred in Israel. Over a 10-year period hundreds of thousands of tons of a mixture of coarse sand and broken shells were extracted for building purposes, causing the shore to recede 4.5–6 m/year as compared with 0.15–0.25 m/year before. In 1964 all shore mining was stopped, and a substitute has been developed. The mining of valuable minerals—rutile, zircon, and associated minerals—has been taking place along the coast of New South Wales, Australia (Barr, 1965; Barr and Atkinson, 1970). In this case, careful operation prevented erosion of the sand. Exhausted areas were re-leveled, the topsoil was returned, and the area was planted, resulting in a stable system of dunes.

HARBORS

Building harbor facilities necessitates either dredging of the sea or construction of breakwaters, piers, and quays. In both cases, material is removed in quantities from one place to another and a new oceanographical and geomorphological situation is created. The disturbed equilibrium reestablishes itself either by depletion or by accumulation of material, or through both processes. Modifying coastlines for harbors is at least six thousand years old (Klimm, 1956). Four thousand years ago many Mediterranean ports already had moles built in deep water. One of the oldest-known cases of a tombolo, formed as a result of human activity, is that of the ancient town of Tyre on the eastern Mediterranean. There, in the 11th century B.C. the Phoenician king Hiram connected two coastal islands to form a sheltered harbor. The currents thus transformed, influencing coastal deposition, formed a tombolo from the mainland to the (unified) islands. Over this tombolo Alexander the Great built in the 4th century B.C. a 60-m wide causeway. In the 2,300 years that have passed since, the tombolo has grown to a full-fledged isthmus of over 1 km width. Another instance where a tombolo has formed thanks to human activities is along the Chesil Bank connecting Portland Island with Dorset County, U.K. In the shadow of the Bank a harbor is now located.

The process of intervention in the natural state for purposes of creating harbors or their improvement has gained considerable impetus over the last 200 years as a direct result of the growth in volumes transported and the size of the vessels. Huge amounts of material, running into the tens of millions of m^3, have been involved (Sherlock, 1922). In a few cases these materials were put to use—as when digging the wet docks of Liverpool, where the material served for manufacturing building bricks (Legget, 1962). In the quest for material to form new "terra firma," domestic and industrial garbage has lately been dumped inside prepared embankments, involving new areas of up to 600 hectares (Kitakyushu, Japan; personal communication by Jean Gottmann, 1976). Other ports of Japan expanding into the sea by artificial islands and dumping of soil are Nagoya, Osaka, Tokyo, etc. This practice also solves the problem of the growing scarcity of suitable dumping areas for industrial waste, such as slags from the steel industry.

The construction of breakwaters associated with harbors influences the regime of currents, as has been set out above. Sand and shingles will accumulate on one side of the structure, with erosion occurring on its other side. The ancient port of Caesarea (Israel), built in the 1st century A.D., clearly demonstrates these effects. To its immediate south, sand has accumulated to such a degree as to cause a landward penetration of the dunes over an area of 100 km^2 (Reifenberg, 1950/51), whereas to its immediate north a Roman aqueduct, built originally parallel to the coast at a distance of some 80 m, has been obliterated by the sea.

We may conclude by observing that for the construction of harbors with their breakwaters, all possible geomorphological interferences must be studied and known prior to construction. Among the main harbors of the world there are several examples where this dictum has been neglected. In Durban, for example, nearly 600,000 m^3 of sand were required for the restoration of a shore eroded as a result of insufficient advance study.

RECLAMATION OF LAND FROM THE SEA

Introduction

Reclamation of land from the sea is one of the main achievements of those people who live on flat, marshy shores periodically inundated by the tides (Demangeon, 1927). This situation is characteristic of the Northwest European countries, where —in some parts, such as northern Sweden, Finland, and Scotland—a long-term upward movement takes place, and in others a constant sinking occurs. The initial measures taken by the inhabitants against the encroaching sea were purely defensive; the oldest settlements in The Netherlands were located on the "terpen," a "terp" signifying a mound some meters above the high tides, high enough to withstand the invasions of the sea.

In contradistinction to a long-range regression of the North Sea over the last few thousands of years, the last two centuries have seen a small, though significant transgression. Constant defense of the seacoast is therefore a technology which must be practiced in those countries.

The coasts of the North Sea are mainly very shallow, dry during ebb and inundated during high tide. Embankments of a few meters only are necessary to convert these terrains into a permanent terra firma. It may be said (Demangeon, 1927) that as a result of these particular conditions, the nations living on the shores of the North Sea occupy a special place in the history of civilization, having created their countries and defending them daily against the sea, the marshes, and the rivers.

Reclamation of Land from the Sea in The Netherlands

The coast of Belgium and The Netherlands, essentially from Calais to Den Helder and the Frisian Islands, consists of sand deposited in form of bars which subsequently emerged after the end of the Flandrian transgression. These sand dunes, colonized by vegetation, became the natural protection of the lowlands. Without them, most of Flanders and two-thirds of The Netherlands would be flooded. Even the rivers are flowing at a higher elevation than the surrounding terrain, thus requiring protective dikes alongside.

Heavy gales and spring tides are a continuous threat to the low terrains behind the dunes, and constant strengthening of the dikes with sand taken from the sea is required. Moreover, shore improvement has reduced the original coast length of The Netherlands by some 300 km (Vita-Finzi, 1974).

One of the greatest achievements of land reclamation in The Netherlands is that of the Zuiderzee ("Southern Sea"), crowning efforts of many centuries (Fig. 51). Although begun in the 10th century, land reclamation required the establishment of windmills in the 16th century to make drainage of water possible from a depth of 3, and subsequently from 6 meters, the mills operating in series to achieve effective drainage to the sea. The areas reclaimed from the year 1540 onwards are known accurately. The Zuiderzee project, covering today some 226,000 hectares, equals all previous achievements. Starting in the year 1919, a 32-km long dike separating this Zee from the North Sea was constructed over 13 years, when the "Zuider Zee" became the IJsselmeer (Lake of the IJssel, named after the inflowing river). The reclaimed land is up to 5 m below sea level. When the last polder of this project is reclaimed, the total area added will amount to over 610,000 hectares, i.e. more than 6,100 km^2, equal to nearly 19% of the total area of The Netherlands.

The conquest of new land from the sea is bound to continue with the projected reclamation of the Wadden Shallows (Fig. 51) between the mainland and the Frisian Islands. Another ambitious project, catalyzed by the severe inundations in 1953 in the southwestern province of Zeeland, encompasses closing the estuaries of the Meuse and Schelde Rivers (the Deltaplan); this aims primarily at an additional decrease of the coastline, with land reclamation as a secondary factor.

Reclamation of Land from the Sea in the United Kingdom

A considerable area in the east of the United Kingdom, especially in East Anglia and Cambridgeshire, is land reclaimed from the sea. It consists of "upland," lying 6 m and more above sea level, and the lower-lying "fens." The area extends, generally speaking, between Cambridge, Lincoln, and Boston.

Earlier bounds of settlement having been determined by the ups and downs of the sea level, it is only in Roman times that reclamation by the draining of marshes preceded extensive settlement, involving systems of canals and dikes.

Most of the processes and the phenomena described in The Netherlands occurred as well in the U.K. The estuaries of the Humber, the Thames, and the Wash suffered floods by high gales, especially when combined with high tides. The 16th century

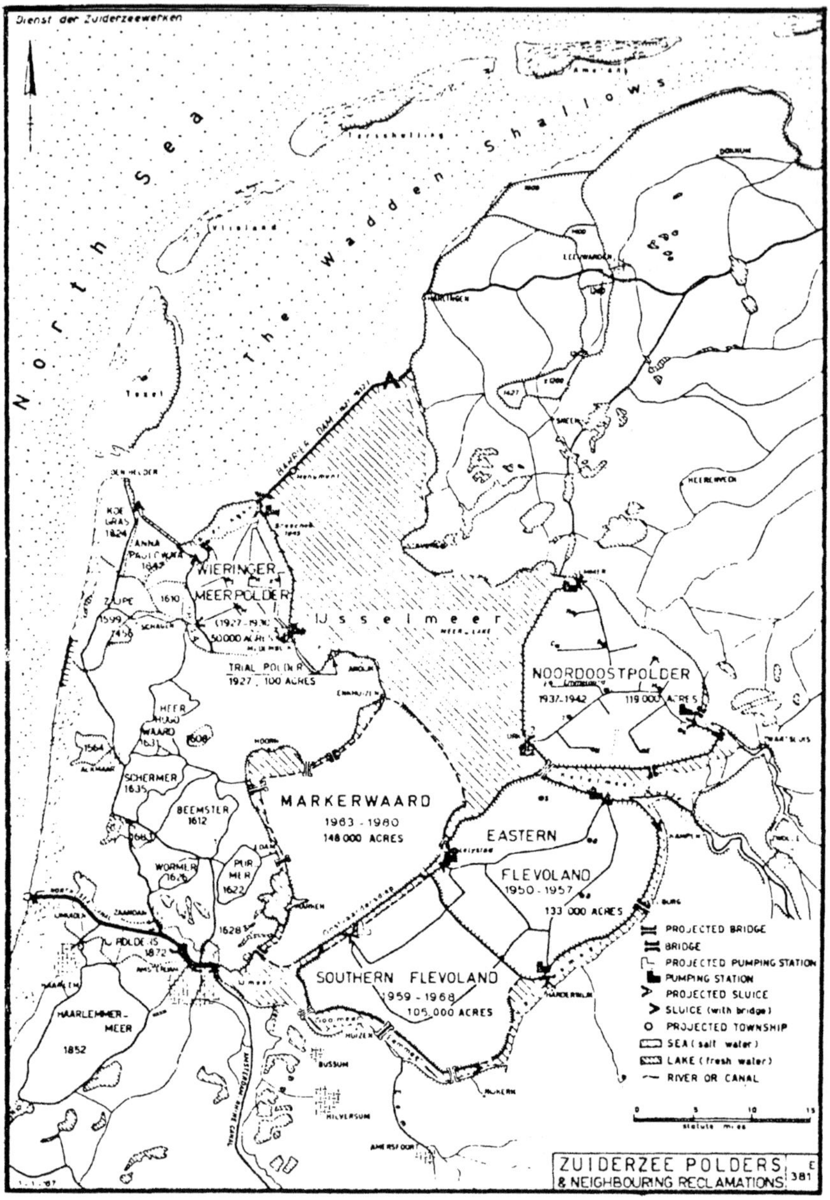

FIG. 51. Zuiderzee and neighboring areas. Names of polders with year of reclamation (Netherlands' Government Information Service, 1967)

saw extensive floods, and it was under King James I that considerable reclamation schemes were initiated. It has been estimated (Jones, 1973), that since 1640, 500,000 hectares of land have been reclaimed. The two most conspicuous floods of this century in East Anglia, in 1953 and in 1975, caused relatively little damage and few victims, proving that earlier lessons had been well learned. Another area reclaimed from the sea is that of the Romney Marshes, Kent, southeast England. This is a low-lying land of 25,000 hectares, acquired over a period from Roman times until the present, but mainly since the 15th century. The present wall is based on the original Roman works.

Similar land reclamation efforts have also taken place in other countries. On France's Atlantic coast, areas in the Bay of St. Michel and in the Bourgneuf region, south of the Loire, have been involved (Verger, 1968). In Italy, the Pontine Marshes, long regarded as a nuisance, were finally drained in the 1920s and added 120,000 hectares to the land.

It may be concluded that reclaiming land from the sea, and defending it against encroachment by the water, is a never-ending task. The Dutch, just to illustrate this point, have a special government branch for this purpose, the Ministry of the State of the Water ("Waterstaat"), that stresses their daily concern with the conquest over the sea.

8

Geomorphology of Settlement

INTRODUCTION

Demography and the Urban Area

Numerous activities of the inhabitants of settlements exert geomorphological influences on the environment: changing the natural area into paved surfaces, such as roofs and streets, thereby disturbing the steady state of the local geomorphological system; extensive use of building materials, excavated from the Earth; extensive use of local and imported water supplies, thus changing the hydrological and hydrographic equilibrium of the environment; creation of a drainage system by means of sewage canals; road building (including access roads); changing the environment beyond the limits of the settlement for recreational needs—parks, swimming pools, ski resorts, etc.

Translated into geomorphological terms, these activities result in the cessation of soil-forming processes in the built-up and paved areas, the formation of new soils in man-made parks and gardens, the reduction of natural hydrographic networks, and the reconstruction and regulation of watersheds (Kotlov, 1972).

This impact on the environment has been evaluated as a most unfavorable one (Leopold, 1968). Various scholars have given widely diverging values to the rates at which erosion of areas, converted to urban uses, occurs. Some evaluations put it at 20,000 to 40,000 times higher than that caused by agricultural and forest use (Wolman, 1974); other estimates are lower. On the other hand, a comparative study of two small watersheds showed a rate of erosion only five times greater in an urban watershed than in a rural one (Leopold, 1968). These greatly different estimates require a closer and more critical evaluation.

One index of the extent of these environmental changes is, of course, the size of the world's population. But one cannot simply take the growth of population as an index of the settlements' impact on the environment, because the rate of growth of the *urban* population is faster than that of the population at large. This trend, characteristic of our century, seems to be worldwide. The world's urban population, which was nearly 3% of the total in 1800 (Weitz, 1973), grew to 6.4% in the following 50 years, and increased further to reach 13.6% in 1900. Fifty years later, in 1950, more than one-quarter of the world's population were town and city dwellers (28.2%); by 1970 the urban population had attained 38.6% of the total, and it is expected to reach 60% by the year 2000 (Kotlov, 1972). The world's population growth from 1.5 milliard in 1900 to more than 4.2 milliard in 1980 reflects not only a relative increase of the urban population but also an enormous absolute increase in the number of human beings needing space for dwellings, roads for communication, water and building materials, all of which are taken from the Earth's resources, and most of them from its surface.

The change in the living habits has not only involved an increase in the urban population but also an increase of size of the particular cities. In 1850, no more than 50 cities in the world had more than 100,000 inhabitants, and only three had more than 1,000,000 (Chandler and Fox, 1974); in 1973 there were more than 1,000 cities with over 100,000 inhabitants and 167 with more than 1,000,000 (Witthauer, 1975).

The spatial-topographical result of this trend is the increase of built-up area in recent years. In Japan, the area under construction was 109.8 km^2 in 1966 and 242.3 km^2 in 1967 (*U.N. Statistical Yearbook*, 1974). In the U.S. the corresponding figure was 566.3 km^2 in 1967, more than two-thirds of it for industrial use. According to our estimate, in 1977 the world's urban area under construction was 9,800 km^2 (*U.N. Statistical Yearbook*, 1978).

However, the size of the built-up area is only one index of the extent of Man's influence on geomorphology; the population density of the city is another. With growing density, more communications, water, and sewage are involved; more roads and railways, more airfields, more parking places—all of them paved and impermeable to water—are required. Cities with a population density of 35 inhabitants/hectare can have open spaces, gardens, and parks, but a city of 1,550 inhabitants/hectare (such as occurs in the centers of the main cities in developed countries) creates quite a different environment. The built-up area of the world is still increasing and has attained more than 10,000,000 hectares (100,000 km^2); 1.5% of the area of the U.S.A. and 10% of the area of the United Kingdom are urban areas (3% for the U.S.A., if airports and roads are included) (McPherson, 1969). These are extents which cannot be ignored by geomorphologists.

Man-made Changes in Urban Topography

The present topography of old cities — Jericho, Jerusalem, Rome, Constantinople, Athens, London, Paris, and others—corresponds all but with that of the primary natural surface of their sites. Each excavation and digging reveals layers of buried streets, houses, and the original bedrock is sometimes several meters below the actual surface. In Jerusalem the present street level near the Western Wall is more than 15 m above the street of the Herodian city, built 1,950 years ago. The surface of central London has been lifted by the accumulated debris of centuries by some 3.5 m

(Brown, 1970). Even in a relatively young city, such as Oxford, which emerged as an urban center only in the late Saxon period (9th century), recent excavations near Hereford College uncovered a rampart of the Anglo-Saxon walls at a level some 3 m below the present street.*

It is the ancient mounds ("tells") of the Near East and other countries, however, which have changed the original topography most. These mounds are the ancient sites of settlements, dating from the third millenium B.C. or even earlier. The destruction of such a settlement—mostly built of adobe or bricks, rarely from stone—and the subsequent rebuilding on top of the debris resulted in the rise of the mound. The original topography—a small elevation in the landscape—rose more than 10–15 m from one occupation to the next. One of the greatest mounds in the United Kingdom is the ancient site of Silbury Hill in Wiltshire. Some 670,000 tons go to make a hill of 30 m height; the original settlement dates from the beginning of the third millenium B.C. Its base measures 2 hectares. In the Near East, however, even greater extents of old cities are known. The city of Hatzor (Upper Galilee, Israel) has an area of 70 hectares, 12 of which are 40 m high (*Encyclopaedia of Archaeology*, 1970).

It is only recently that we have attained the technical capability of changing the topography as found. An ideally located town might occupy a level surface; such, however, is the exception. There is today scarcely a big city which did not change its original topography when developing new quarters; it is this leveling practice which causes most of the erosion in the construction phase of an urban settlement.

Modern urban growth intervenes in the original topography in three ways: by cutting down elements from the relief, thus deleting from the topography; by building ramps and surfaces above the original relief, thereby adding to the topography; and by reclaiming building areas from an originally unstable environment, such as marshes, swamps, and sea shores. In this way 9.5 hectares of new land were added to Manhattan Island (Legget, 1973); Milwaukee reclaimed 15 hectares from Lake Michigan. Japan established in 1970 a national scheme to reclaim 47,000 hectares around its major ports. Building on slopes of up to 25% is now routine practice in countries where urban areas are scarce and expensive. Stripping the alluvial soil from a valley and transporting it to a residential rocky hill top, thereby creating gardens, is the only way of improving the landscape in a karstic country. The two modern, expanding towns of Rio de Janeiro and Hong Kong illustrate well the rapid change of the original topography. The site of both cities is characterized by the favorable circumstance of a good and deep harbor, but—on the other hand—by the narrowness of the coastal plain sandwiched between hard, rocky, and steep hills and the sea. In both cities reclamation of land, both from the sea and from the hill slopes, has taken place on a considerable scale.

The most striking features of the Rio de Janeiro surroundings include the considerable Guanabana Bay and the drastic differences in elevation due to the mountains, hills, and escarpment (Morris and Pyle, 1971). Three east-west ranges of mountains penetrate the city, dividing it into a southern and northern part. Until

* Prof. Jean Gottmann, personal communication, 1975.

the 1950s passage from one part to another was quite difficult, and required the use of the sea shore or a gap in the mountain range. In the 1960s the situation was remedied by excavation and removal of the smaller hills, boring tunnels through the granitic "sugar cones," thereby opening up the previously remote areas to development, such as the Copacabana quarter. The removed material so obtained was usefully employed to extend the limited area of flatlands adjoining the sea by partly filling the bays of Botafogo and Flamengo. The double operation of removal and filling aimed at achieving the lowest possible gradient of the new topography. Although the human intervention aimed at serving urban transportation, we may consider the result to be similar to that which long-term natural processes would have achieved. In this case, Man's intervention could be viewed—geomorphologically speaking—as almost harmless. The type of climate, with heavy seasonal rains, required, of course, runoff control by appropriate means (ditches, culverts, planting lawns and trees, etc.).

The case of Hong Kong is perhaps even more striking than that of Rio de Janeiro. Control of the surface growth is dictated not only by environmental factors but by political ones as well—the proximity of the frontier of China and the need for containing the ever-growing population. The circumstances favoring a good harbor

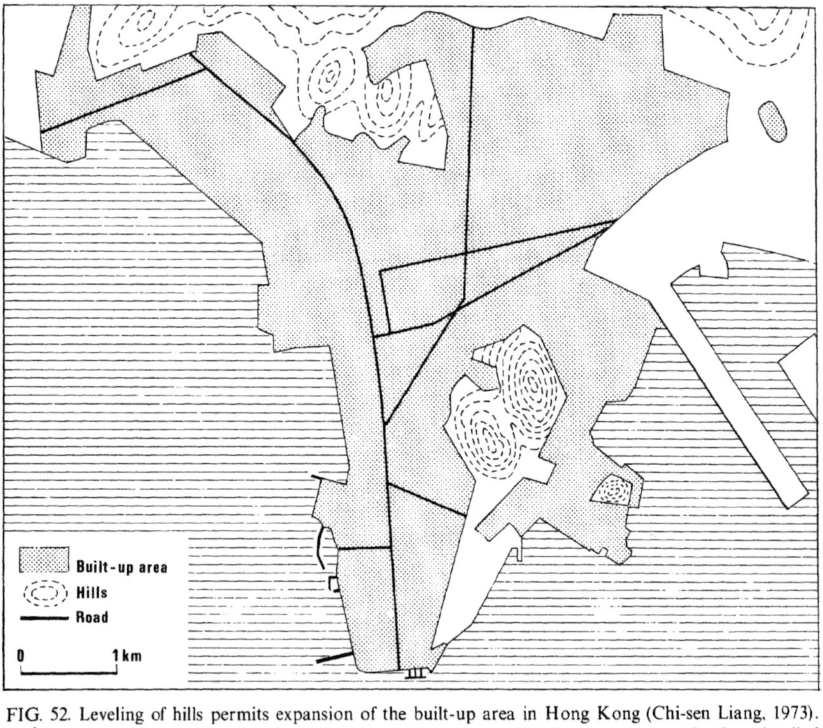

FIG. 52. Leveling of hills permits expansion of the built-up area in Hong Kong (Chi-sen Liang, 1973). The contour lines indicate the tops of hills; the area around these hills that was put to use has been levelled

are balanced by the natural challenges of the site. The urban area of 1,036 km^2 is very rocky, with a high energy relief (Chi-sen Liang, 1973). On 905 km^2 (about 88% of the area) only 5% of the inhabitants are living; the other 95% of the population are squeezed into the remaining 12% of the area. The increase of the population was so tremendous—from 1.6 million in 1942 to 4.1 million in 1972—that the urban density had risen by 1972 to 200,000 inhabitants/km^2, the highest in the world. Small wonder that every possibility of increasing available area has been followed up, either for facilities which need a plane surface (airport, harbor) or for housing. The additional land has been reclaimed either from the sea (the depth of which, near the coast, is quite considerable) or by leveling the slopes of the hills (Fig. 52). The development of new land is favored both by government planning and by investment policy.

In the period since 1851, over more than a century and a quarter, the reclamation of land has been one of the main development trends of Hong Kong. The total area of reclaimed land before World War II was 36 hectares; since World War II, more than 405 hectares have been added, amounting to 25% of the present built-up area (1,600 hectares). This hectic intervention into topographical elements has sometimes had a disastrous effect on the stability of slopes, with landslides known to occur in the built-up area (p. 127).

Stages in the Environmental Impact of Built-up Areas

Modern settlement is, as we have shown above, mostly of an urban character; farms and villages have the same environmental impact as a town, only on a smaller scale. The effects of such settlement should be differentiated according to the different stages of development. The effects consist essentially of removal of the natural vegetation, restriction of the natural drainage, and accelerated import and export of water into and from the urban areas, altering the original topography, hydrography and hydrology, resulting in a profound disturbance of the ecosystems. These effects should be discussed for the three separate stages (Savini and Kammerer, 1961; Wolman, 1967; McPherson, 1969) as follows: (1) the transition from the pre-urban to the early-urban stage, or the "sub-urban" period; (2) the transition from the early-urban stage to the middle-urban stage, or the "construction" period; (3) the transition from the middle-urban to the late-urban stage, or the "developed-urban" period (Figs. 53, 54).

(1) In this stage land use is still predominantly rural and any building activities generally are in equilibrium with the geomorphological forces. At the beginning of this stage, the landscape is mainly an agricultural one. Increasing economic activity results in removal of trees and other vegetation and in the construction of scattered houses equipped with only limited water and sewage facilities. The effects of this stage are mainly a decrease in evapotranspiration resulting from the clearing of vegetation and an increase in stormflows and sediment deposition in the streams.

(2) This stage is characterized by its construction activity. Construction areas are increasingly exposed to climatic influences, whereas measures taken to lessen the erosion manifest themselves only partially. In fact, when speaking of environmental damage caused by urban land use, it is mainly this stage which we have in mind. It

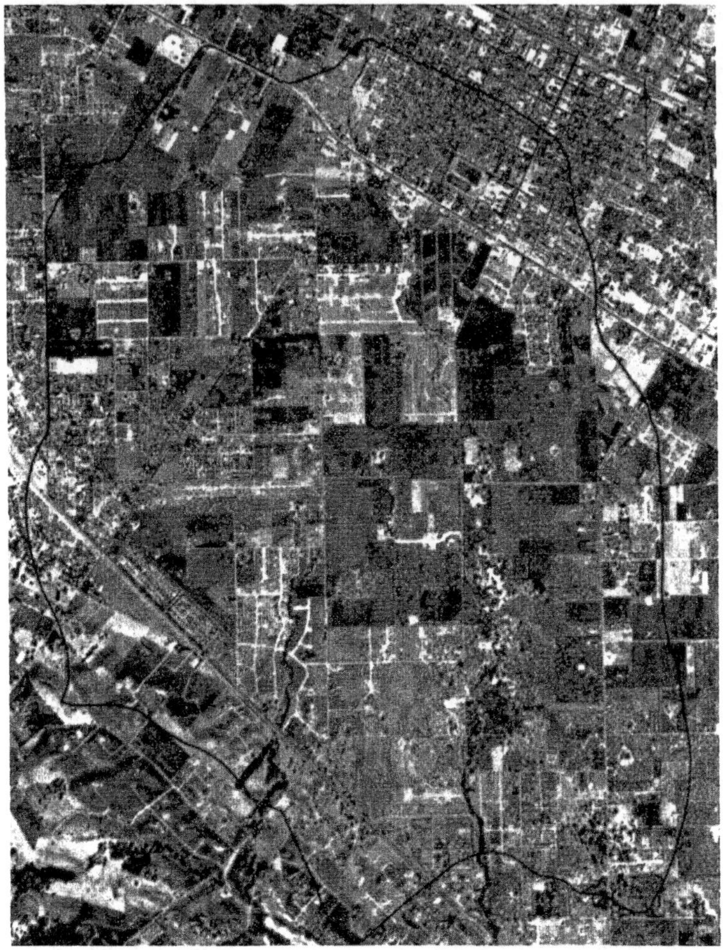

FIG. 53. Urban growth in Santa Clara County, California, Permanent Creek project. The project area in 1948 (Harris and Rantz, 1964)

is true that such a view is partially an oversimplification of the problem. Still, this is the stage of bulldozing and cutting, removal of the topsoil, combined with the building of houses, the paving of streets, and the construction of culverts. The immediate effects are increased erosion on the construction site and aggradation downstream. Existing streams are eliminated by reshaping the terrain; the area of paved surfaces is increased, whereas any new system of drainage has not yet been wholly constructed. All this results in a decrease of infiltration and, evidently, in increased danger of floods, carrying considerable amounts of silt, and clogging the culverts.

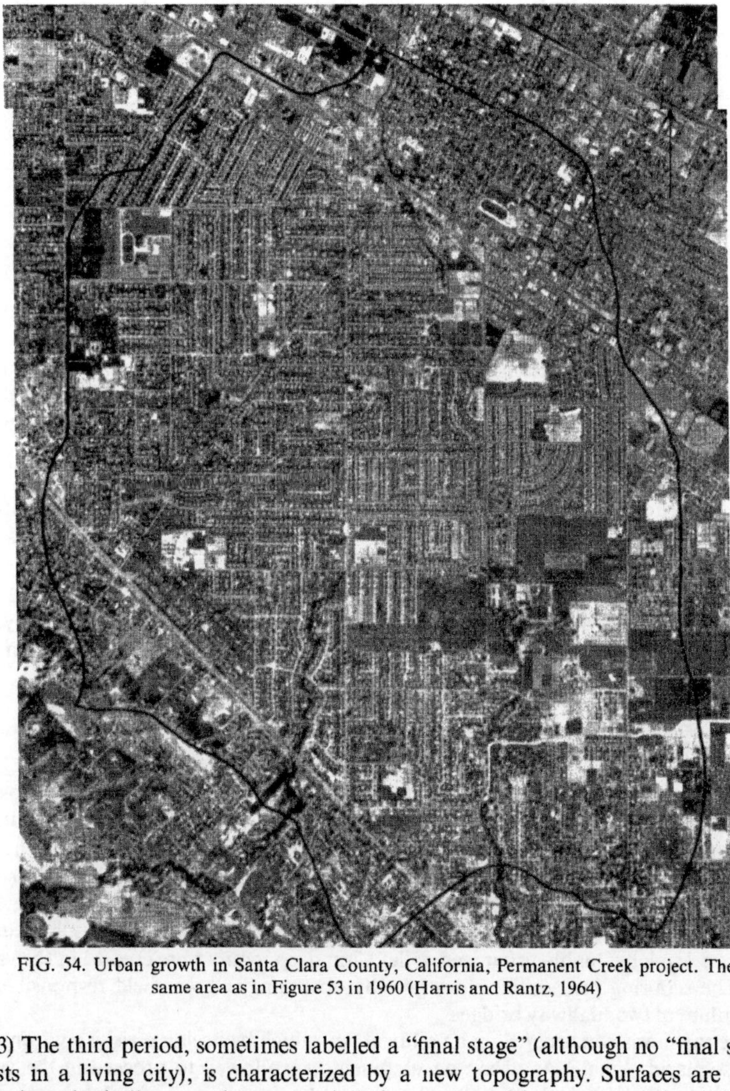

FIG. 54. Urban growth in Santa Clara County, California, Permanent Creek project. The same area as in Figure 53 in 1960 (Harris and Rantz, 1964)

(3) The third period, sometimes labelled a "final stage" (although no "final stage" exists in a living city), is characterized by a new topography. Surfaces are either paved or else built up, and are essentially impervious with the water being led away by sewers, gutters, or culverts. Even in the case of lawns and gardens, the new vegetation prevents exposure of the bare soil. This is a stage where a quarter is completely urbanized, where the runoff is discharged into streams outside the town area, where no other runoff channels exist than the artificial drainage system, and where sewage is fully treated. In this stage, the flood peaks in the natural channels downstream

of the town become higher, because of the higher runoff from the paved surfaces; sedimentation decreases, as the runoff from the urban area carries less solids. Infiltration into the groundwater table is reduced, and flood damage to downstream areas is generally higher.

MORPHOLOGY OF THE "CONSTRUCTION" PERIOD

Runoff and Stream Discharge

Urban development changes two of the basic elements — the infiltration rate and the lag time — which determine the magnitude and timing of the volume and the peak of the flood hydrograph (Carter, 1961):

1. The average infiltration rate is *decreased* as a result of the construction of an impervious area consisting of roofs and paved streets; on the other hand, in those areas where the original vegetation has been cleared away but no buildings or streets as yet constructed, the infiltration rate may even *increase.*

2. The lag time, defined as the time lag between the peak of the rainfall and that of the resulting flood, is *decreased* as a result of the absence of natural vegetation; later, this effect becomes even greater because of the installation of storm sewers and the improvement of the principal channels.

Studies of Yorke and Davis (1971) on Bel Pre Creek, Montgomery County, Maryland, by comparing an area before and during construction periods in the years 1963–1967 revealed an increase of 30% in the runoff, and of 1,400% in the sediment, caused by construction on 15% of the watershed area.

Soil Erosion

Urbanization of any area may result, as shown above, in increased soil erosion. Often, however, erosion results from natural causes, reinforced by human intervention. A town could, for example, be located on an unstable alluvial site, thereby contributing to erosive forces, but not be wholly responsible for the release of eroded material itself. This seems actually to have been the case in the extensive scour which occurred in Tujunga Wash, southern California (Scott, 1973). Huge floods occurred in this 5-km long, highly urbanized valley; 7-m deep scours were created in the valley fill. The scouring process moved gradually upstream and was held responsible for the failure of two highway bridges.

The town was constructed on an alluvial fan, which by itself is a risky environment in a semiarid climate. As urban development was allowed to proceed in the valley before completion of adequate flood measures (sewage systems, bank management, etc.), we may consider that a man-made component contributed to the erosion. Leighton (1972) assumes that most destructive landslides in the populated areas of California are related to human activity. Landslides occurring subsequent to construction activities have been associated both with the graded as well as with the natural segments of hillsides and represent either the rejuvenation of old landslide scars or a development of new ones. A landslide map of south California in 1972 showed 4,000 landslides; about 50% of the landslides originating between 1966 and 1972 are estimated to have been caused by constructional activity on steep slopes consisting of unstable rock material.

Sedimentation

The process of sedimentation in the early urban stage has been studied, inter alia, by Guy and Ferguson (1970) in Kensington, Maryland, and by Maderey (1974) in Mexico City. These two studies confirm the increase of sediment load in the streams in this stage of urbanization. Guy and Ferguson (1970) investigated a construction site on a watershed of 24 hectares. During the period from July 1959 to July 1962, 89 detached houses were constructed on 8 hectares in the upper part of the watershed. During the same period, 1,127 ton/hectare of material were lost from the construction area, each lot being exposed to erosion for periods from 8 to 24 months (Fig. 55). In the storm runoff, sediment concentration reached 44,000 mg/liter, with one observed peak of 90,000 mg/liter, for a flow of 2.35 m^3/sec. At the end of the construction period the stream channel became wider than it had been before, as the result of both increased runoff and increased coarse bed load.

The effect of forest clearing to enlarge an urban area is shown in Figure 56. This is a graph of the discharge and bed load in the area of Mexico City during the years 1961–1971 (Maderey, 1974). The western part of this city grows towards the top of the slopes of Mexico Valley, demanding extended clearing of the forest. Because of the seasonal character of the rainfall, and—not less important—because of the

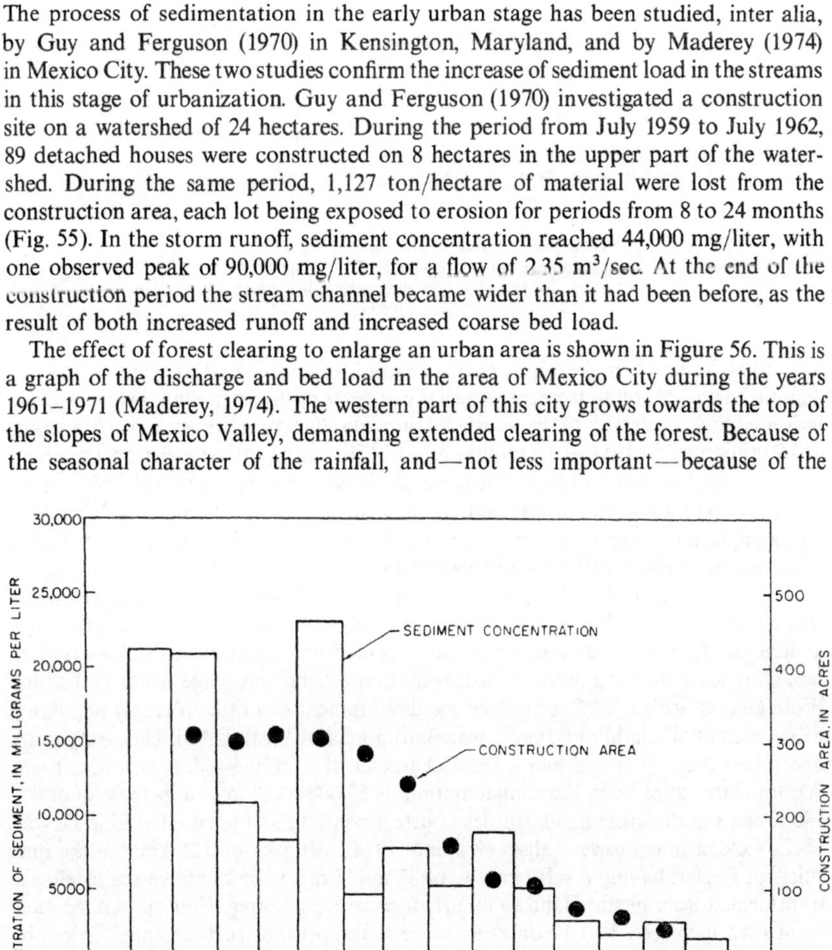

FIG. 55. Variations in the mean stream sediment concentration and urban construction area (Guy and Ferguson, 1970)

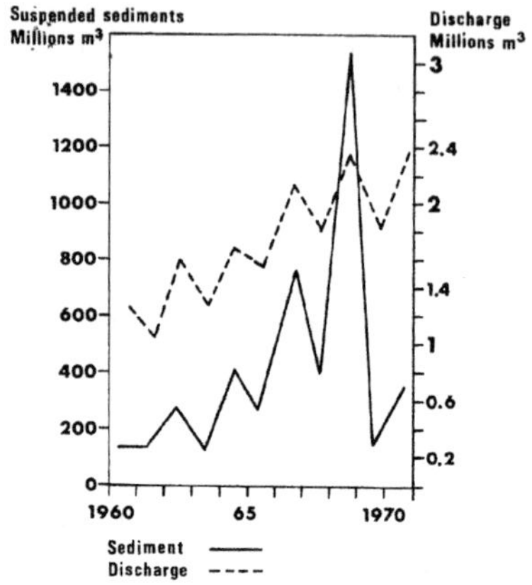

FIG. 56. The influence of forest clearing for the urban development of Mexico City on the sediment yields and runoff discharge, 1961–1971 (Maderey, 1974)

instability inherent in the high relief formation of the cleared area, there exists a definite hazard of soil movement, especially in years of heavy rainfall. Measurements indicated that, as a result of the forest clearing, runoff increased from a yearly average of 1.2 million m^3 in 1961 to 2.3 million m^3 in 1971; during the same period, the urban area increased by some 13 km^2. The increase of the sediment load was highly irregular, from 200 million m^3 to 400 million m^3 and then to 1,500 million m^3. In 1969, for example, an extraordinary increase of sediment load could be directly correlated with the heavy deforestation taking place then.

It can be assumed that the period of "construction" is one of the *worst man-made erosions ever possible*; but as this period is limited, its data cannot represent an average annual rate for the overall sedimentation as caused by urbanization. Still, an evaluation of at least the magnitude of sedimentation during this stage could be helpful. Wolman and Schick (1967) examined the development of a rather densely populated urban area (400 inhabitants/km^2) near Baltimore and Washington, D.C. Fifty years ago, when the region still had a rural character, the annual sediment amount was 400 ton/km^2; at present, the annual amount is 55,000 ton/km^2, an increase of nearly 140 times. On the other hand, studies of urban construction areas effected in Devon, U.K., yielded much lower values (Walling, 1974). An area of 0.26 km^2 on the outskirts of Exeter, having a relief energy of 83 m (72 m to 155 m above sea level) was transformed from pasture land to an urban construction area. Here suspended sediment load increased 5 to 10 times its value at the preconstruction stage. This is by itself quite a spectacular change, although it is smaller than that reported above.

It seems that there are more complicated relationships governing soil erosion than removal of natural vegetation only; one should take into account also the different degrees of soil conservation during the construction period. A variability in the rates

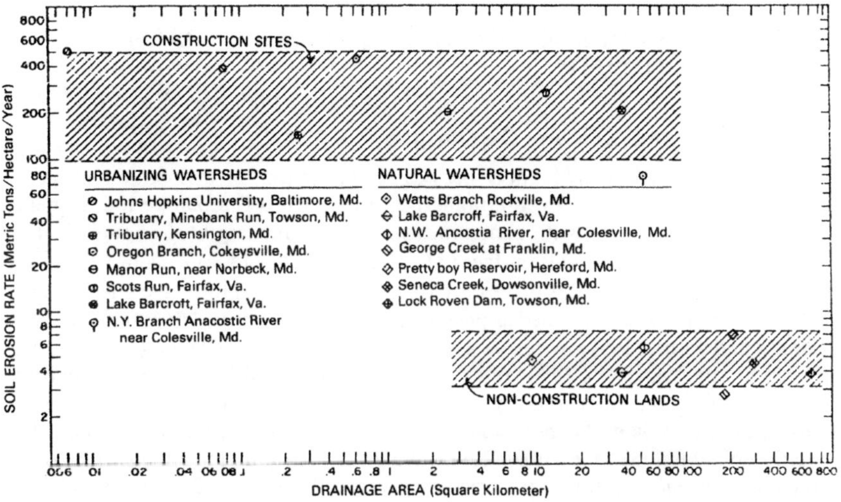

FIG. 57. Soil erosion rates from construction and nonconstruction areas (Chen, 1974)

of erosion has been shown by Guy (1974), who examined a construction site at Restan, Virginia. On a site of 200,000 m^2 the rainfall of 1972—1,498 mm—generated a yield corresponding to 4.4 kg/m^2; on the same site, the rainfall of a year later—1,450 mm — generated a yield corresponding to 2.3 kg/m^2 only. This drastic decrease by nearly 50% in the rate of erosion should be attributed to erosion control measures taken and shows the short duration of the erosion during the construction stage.

To establish an estimate of the sediment yield during the construction stage, Chen (1974, cf. Fig. 57) proposed the following equation:

$$Q_s = R\left[\left(Q_c\frac{A_c}{A}\right) + q\cdot\left(1 - \frac{A_c}{A}\right)\right]A$$

where Q_s = sediment yield; Q_c = soil erosion rate from construction sites; q = soil erosion rate from nonconstruction sites; A_c = area of the construction site; A = area of the basin in which urban construction occurs; and R = sediment delivery ratio = % of sediment delivered from the erosion source to any particular location in a basin. All other factors being known or measurable, the sediment delivery ratio, and hence the total sediment delivery, can be calculated. Chen's evaluation leads to a delivery of about 300 ton/hectare/year (30 kg/m^2/year) on a construction area without erosion control practice, decreasing—according to the degree of control practice—down to 5 ton/hectare/year (0.5 kg/m^2/year).

GEOMORPHOLOGY OF THE DEVELOPED URBAN ENVIRONMENT

Runoff and Stream Discharges

In contrast to the construction period, characterized by denudation of urban areas from their protective vegetation cover, the developed urban period is typified by the sealing of the soil with impervious surfaces. Estimates put the figure of runoff from completely impervious surfaces at 75%, compared to 30% from natural untouched surfaces. The effects of the developed urban topography on the intensification of the runoff have been studied, especially in the U.S. (Seaburn, 1969; Sawyer, 1963; Harris and Rantz, 1964; Wolman, 1967; Anderson, 1970), in Japan (Miyata, 1969; Kinosita and Yamazaki, 1974) and in the U.K. Harris and Rantz (1964) as early as 1964 described the correlation between the increase in runoff and the increase of urbanization. In Santa Clara County, California, they compared the 1945 and 1958 *outflow* of runoff (i.e., the streamflow leaving the area plus the channel seepage in the area) and *inflow* (rainfall plus incoming streamflow). The rate found was 1.18 in 1945 and 1.70 in 1958. The increase was attributed to the increase in impervious surfaces from 5.7% of the municipal area to 19% resulting from the conversion of orchards to dwellings, each house adding some 300 m^2.

The area of totally impervious surfaces increases with intensification of the urbanization. In a suburban type of settlement, even after completion, the impervious surface takes no more than 11.5–12% of the total area (Carter, 1961). Plots of 1,400 m^2, as are common, have impervious surfaces of 25%. In downtown quarters, the impervious surface may amount up to 90% of the total area. Brater and Sangal (1969) established a relationship between the increase of density of urban population and the impervious areas. When the density of population varies from 200 to 2,800 per square kilometer, the impervious surface increases from 1 to 10%. The limited infiltration into the soil causes the rate of infiltration in a developed urban environment to drop to one-sixth of its value in a natural environment, such as a forest (Lindh, 1972).

The second impact on runoff is caused by the system of area drainage by sewers. This system causes a reduced lag time of the peak flood and an altered hydrograph. In a completely sewered basin, the lag can be reduced by 85% (McPherson, 1969). These two effects combine, and where man-made drainage system meets the natural, unchanged one, increased flooding occurs (Strahler and Strahler, 1973). Labeling such floods as "man-made" describes the situation truly. Teller (1968) characterizes these floods by saying "The occurrence of flow is of such a magnitude that the natural banks of the waterway are overtopped and water flows over the low-lying adjacent areas."

The average runoff hydrograph is strikingly different from the natural one (Fig. 58; Lindh, 1972); it is much more concentrated in time and has a steep peak. But there is also an increase in the volume of runoff: Anderson (1970) states that a complete man-made channel system, having a completely impervious surface, may in-

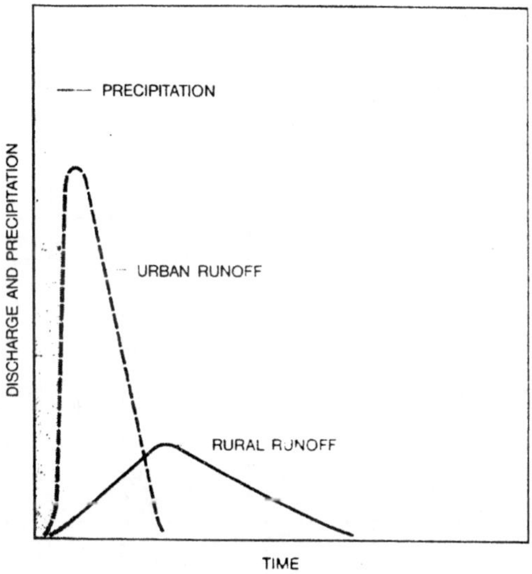

FIG. 58. Differences between rural and urban hydrograph (Lindh, 1972)

crease the floods volume downstream by a factor of 8. Martens (1968) states that a developed urban area could double the discharge of a flood, such as might occur every 20 years. This flood could raise the stream level to a height endangering a whole city.

A study of the relation between the pre-urban and urban dimensions of runoff was made by Seaburn (1969) on Long Island counties, U.S.A. The area under study included the 40-km^2 basin of East Meadow Creek. Its urbanized part, "Hampstead Subarea," covers about one-third of this watershed, some 1,400 hectares.

Here, as elsewhere, construction of impervious surfaces and drainage systems formed the major physical change associated with urbanization. The extent of the sewered area was 230 hectares in 1943 and gradually increased to 1,457 hectares in 1962. Runoff volumes were as follows:

Period	Area drained by sewers (hectares)	Increase over initial value (%)	Runoff average yield (m^3)	Increase over initial value (%)
1937–43	230	—	1,134,000	—
1944–51	291	27	1,443,000	27
1952–59	1,327	480	2,714,000	140
1960–62	1,457	530	4,194,000	270

The great increase in 1952–59 coincided with rapid housing development. The increase of runoff volume is clearly related to the increase of the sewered area, although no direct proportionality holds (Fig. 59).

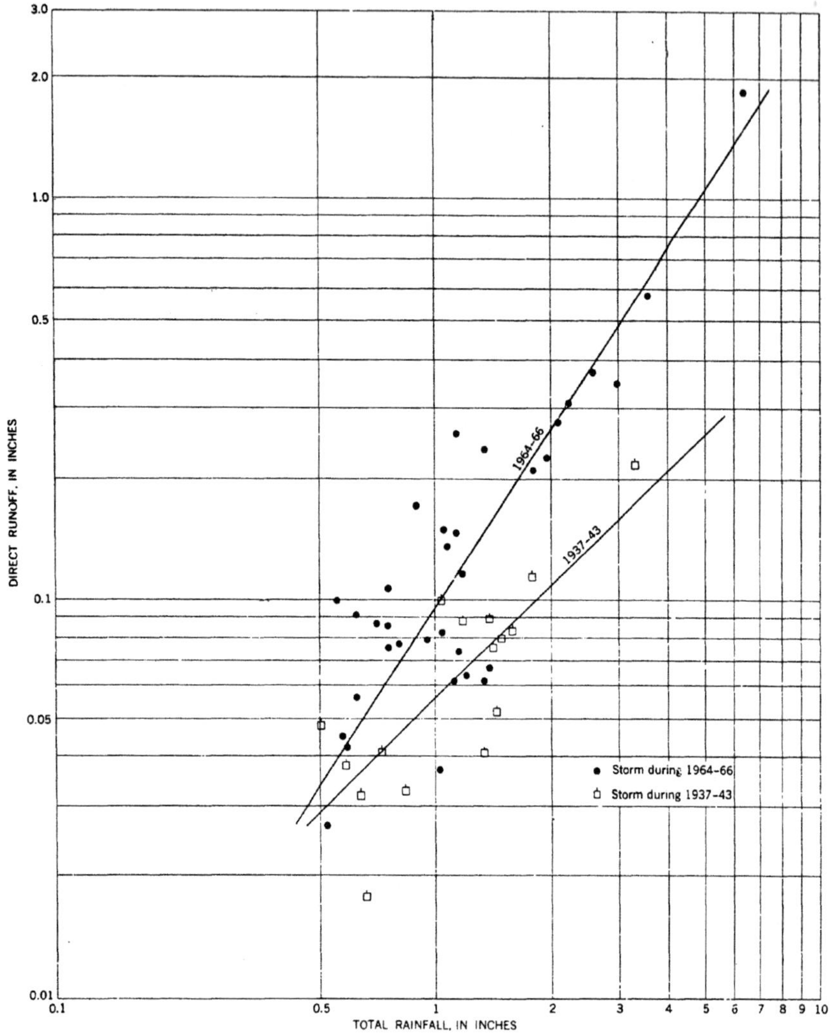

FIG. 59. Rainfall-runoff relationship in the Hampstead area for pre-urban (1937–43) and urban (1964–66) conditions (Seaburn, 1969)

The study of Seaburn showed that concurrently with increased runoff volumes, peak discharges increased up to 2.5 times their value before sewerage had been installed. Planners and engineers would do well to heed this fact when designing sewage projects of new quarters and build a drainage system based not on the discharge of the pre-urban stage, but on a multiple of it of at least 2.5 times if not more.

Erosion and Sedimentation

Mass movements

Man-made slopes represent a most unstable equilibrium, especially when they are steep, and the environment combines heavy rainfall with an unstable bedrock. Hong Kong, where human intervention has changed the topography so considerably, has suffered from landslides triggered by cutting for roads and buildings, even though slopes were extensively strengthened by vegetation or walls (So, 1971).

The typical Hong Kong summer is hot and wet, with heavy monsoon rains. On June 12, 1966, 400 mm of rainfall were recorded, equaling an average month's precipitation. A series of landslides, starting the same day, caused the death of 64 persons, making 2,500 homeless, and necessitating the evacuation of 8,000 more.

In all, 588 events, including large-scale and deep-seated landslides, rotational shearing and slumping, and 144 smaller surface washouts were counted. The examination showed that of these smaller washouts one-sixth had natural causes, and five-sixths were caused directly by human activities. Of the landslides, one-third had natural and two-thirds had man-made causes. The landslides occurred even on slopes controlled by landscaping and covered with vegetation. It seems that protecting the steep, man-made slopes by vegetation alone was not effective; because of bedrock weathering, the vegetation clung only to the upper soil layer. The landslides were the result of a natural catastrophe—heavy rainfall, triggered by building on steep, unstable slopes on a weathered bedrock.

Sedimentation and channe erosion

The most conspicuous fact concerning sedimentation and siltation in a developed urban environment is, that in spite of increasing runoff compared with the pre-urban stage, the sediment yield *decreases.* The heavy sediment load of the construction period is thus only a transitory albeit an intensive one. Thompson (1970) goes so far as to the state that in stabilized urban areas the erosion rates are in general well below those of agricultural areas having similar topographical and climatic characteristics, 0.20–0.24 ton/hectares/year in the former as against 0.21–0.80 ton/hectares/year in the latter.

These two qualities of the discharge from developed urban area—an *increased* runoff and a *decreased* bedload—confer a condition of nonequilibrium on the stream. As a result, where the urban solid-wall, channel-confined stream meets the natural conduit, increased channel erosion and channel widening occur (Wolman, 1967). Observations by Douglas (1976) confirm this rule. The same effect was observed by Mosley (1975) in the U.K.: the Bollin River in the Midlands, formerly a meandering stream, changed over some 30 years to become a straighter, quicker waterway, the lobes having disappeared. Mosley attributes these changes to the increased discharge of developing Manchester and its suburbs.

Hydrology of Developed Urban Environment

The hydrological cycle, operating as it does on a worldwide scale, constitutes one of the greatest systems of physics; in terms of orthodox earth sciences, man can have only a local and limited temporary effect on it (Legget, 1973). This "limited" and "temporary" effect may be considerable, sometimes even crucial, for local environmental conditions, especially if the changes introduced are intensive ones. This is the case of a dense urban society. The change from natural to urban hydrological system is shown in Figures 60 and 61, worked out by Lindh (1972). The most important changes induced by urban society into natural hydrological systems are as follows:

1. Import of water for urban needs from other watersheds.
2. Export of water by sewage and, as a result, a local decrease of infiltration.
3. Decrease of infiltration by limiting the pervious surfaces, resulting in depletion of groundwater resources.
4. Depletion of subsurface resources by intensive drawing for urban purposes; this effect is, however, sometimes partly compensated by leakage of the sewage system.

When water is imported from a distance—such as is the case of Paris, New York, Singapore, and other towns — the effect of urbanization is considerable. Distant

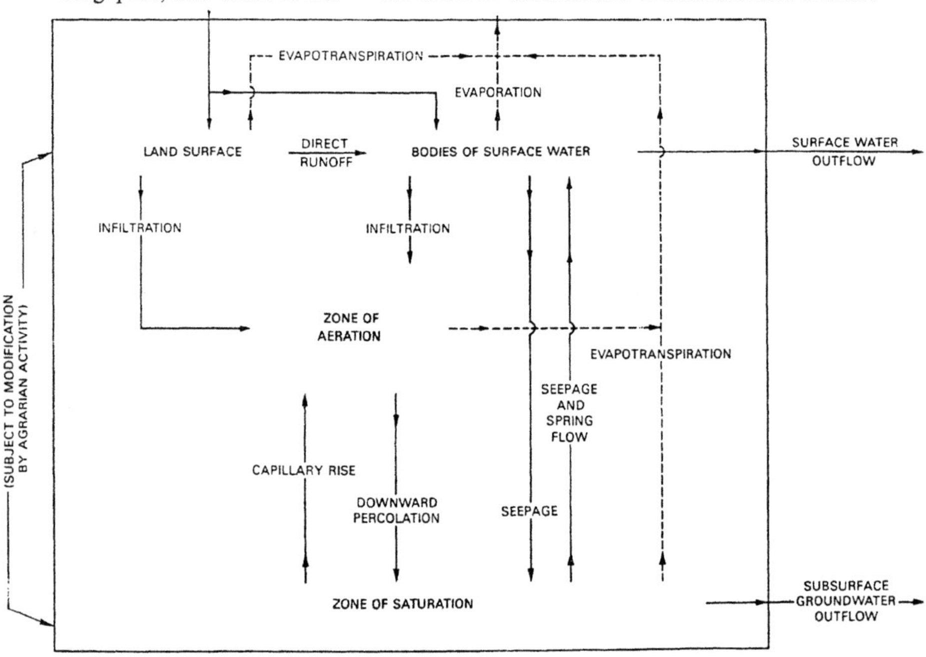

FIG. 60. The "pre-urban" hydrological cycle (Lindh, 1972)

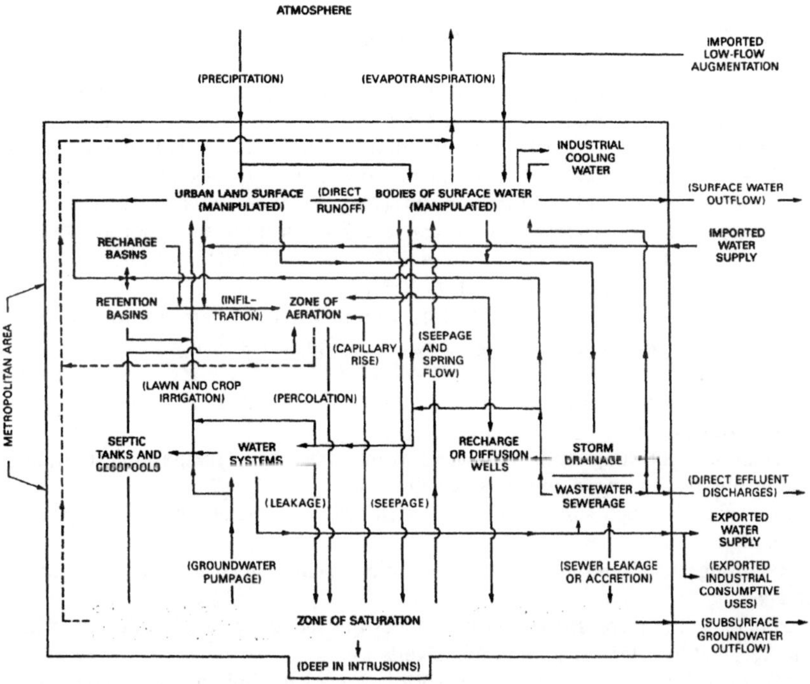

FIG. 61. The "urban" hydrological cycle (Lindh, 1972)

watersheds are deprived partially of their water and the hydrological equilibrium is disturbed. Not only big metropolitan concentrations but smaller communities also may effect the watersheds from which they are drawing their water.

By the joint action of surface sealing to infiltrate water and the intense pumping for urban needs, groundwater resources are exhausted. Some very spectacular drops of the groundwater level in the great urban areas have occurred. In the Paris Basin, the level of the water table dropped about 80 meters in the second half of the last and in the first half of this century (Legget, 1962). In the Chicago district, the groundwater level drops on an average 1.2 m per year; in the London Basin, in spite of its situation around the Thames River, the level of groundwater has dropped since 1878 at a rate of almost 45 cm a year.

This trend of exhausting water resources can be countered by leading the storm waters into groundwater reservoirs instead of into the natural channels outside urban areas. At Nottingham, U.K., water has been fed back (into the Lias limestone layers) since 1890. It is, however, only since the 1940s that urban development has made groundwater replacement urgent. In the Nassau and Suffolk Counties, Long Island, the method has been used of discharging storm water into "recharge basins" from which the water reaches the groundwater level (Seaburn, 1969).

GEOMORPHOLOGICAL MANAGEMENT OF URBAN AREAS

No adequate and accurate figures on the urban-triggered sedimentation and erosion are as yet available; there exist only tentative estimates. The yearly global rate of erosion created by urban construction (= area under actual building multiplied by the rather low estimate of a rate of erosion of 18 ton/hectare/year, cf. Chen, 1974) has been estimated as 17 million tons (Nir, 1977). It is obvious, therefore, that geomorphological management of urban areas constitutes an important task of town builders. Powell et al. (1970) outlined the means needed to achieve better management of the urban geomorphologic processes:

Fitting development to climatic factors, to topographical elements, to soil conditions, and to vegetational cover, so as to disturb previous steady states as little as possible.

Reducing as much as possible the duration and the extension of denuding construction areas.

Retaining, protecting, and, wherever possible, improving the natural vegetation, minimizing its clearing as much as possible.

Covering disturbed soils, especially on slopes, with mulch, grass, or any other vegetation.

Retarding runoff and erosion by mechanical means, such as banking, terracing, and paving.

Managing the increased runoff by ditches and concrete canals, adequately distributed, and by extended sewers and culverts.

Means such as these may limit the erosion process to a minimum and convert the urban area into an integral, almost nondisturbing element of the physical environment.

9

In Conclusion

METHODS OF RESEARCH

A science is distinguished by its methodology and its paradigms. The presentation of the paradigms of anthropogeomorphology, as of other sciences, often depends on the quality and sophistication of the methods. On the other hand, the methodology of anthropogeomorphology depends on the broadness of approaches, which should include the following:

A historical approach, to investigate Man's intervention in the land forms, land uses and geomorphological processes.

A socioeconomic approach, to investigate the dynamics of Man's activity, which depends on his economic, social, and iconographic qualities.

A geomorphological approach, to investigate the rate and extent of the geomorphological processes observed.

A planner's approach, to unify the various viewpoints.

The stress given to each of these approaches will differ, depending on the actual situation studied. Ideally, every physical construction should be preceded by a preliminary study on the rate of geomorphological processes, to restrict the harm caused by Man's intervention to a minimum and, at the same time, to integrate the construction into the natural system.

Methods of research might be classified as follows:

Investigation of an area "before" and "after" Man's intervention

This methods fits best separate interventions over a limited period, such as a change in land use, construction of a road or a building, etc. This method is the oldest in anthropogeomorphological research, easily applicable in future, and—with the aid of air photographs—useful to the study of happenings in the past. Most of the case studies quoted in this book fall within this type.

Investigation by simulation of an anthropogeomorphological process

This method consists of isolating a certain process, and its investigation in an "artificial" environment. Examples where it would apply include the clearing of vegetation, a distinct type of ploughing, and others. When compared with the actual processes, this method introduces less "noise," since it investigates only one, or at most a few, of the factors of the environmental system. The advantage of this method is obvious; it results in measurable and reproducible values of the rate of the process investigated. The integrated role, however, of the process in its environmental system is not revealed by knowing this rate.

Investigation of completely man-made environments

Examples of such investigations include the measurements of stream discharges from urban and constructional culverts, measurements of the volume of sediments trapped by a new groyne or a new wave breaker, measurement of the silt content of a river in a watershed which has been drained, irrigated, or canalized, and others. These investigations give a quantitative estimate of the extent of the processes. The integration of such rate measurements into a complete description of a process remains (as above) problematic. Studies in which the whole area encompassing the investigated process is included may sometimes assist in presenting an integrated view.

Investigation of the socioeconomic behavior, as it affects the rate of anthropogeomorphological processes

When an investigation in one of the three mentioned ways results in a quantitative rate for any process, the outcome should be held valid under the given conditions of place and time only. The reason for this limitation lies in the variability of the human factor with time and with place. The socioeconomic behavior of the people involved is therefore a necessary subject of study within any anthropogeomorphological investigation.

RATES OF ANTHROPOGEOMORPHOLOGICAL PROCESSES

Obviously, the rate of processes as they are influenced by human intervention is of importance for evaluation of the impact of those processes on our surroundings. The difficulty lies, however, in determining the initial natural conditions and rates (Leopold, 1956). Trying to base ourselves in this book on original trustworthy reports, we found out with Lewin (1975) that they "were often confused and, wordily, indigestible." Take, for example, the effect of Man on erosion, where the estimates (expressed in $m^3/km^2/year$) vary from 1,500 to 85,000 (Brown, 1970), with the natural background erosion estimated (in the same units) from 12 to 1,500. The wide ranges would preclude any valid deductions as to Man's influence.

On top of this incertitude, the disparities in the quality of the statistical material used obviously add to the general uncertainty as to the quality of the rates of the processes as obtained.

Forest Clearing

About 1% of the world's forests, estimated in 1976 at 4,145 million hectares, are annually exploited (FAO, 1979). Estimates of annual erosive rates are between 2.5 and 25 tons/hectare (Douglas, 1967; Bennet, 1939), giving a total erosion of 110–1,100 million tons/year. Under tropical and subarctic conditions, where most lumbering takes place, the use of heavy tractors and the large-scale employment of dirt roads will tend to increase the erosion. We suggest a value of 800 million tons/year for this branch of economic activity.

Grazing

The total area of pasture land in 1976 has been estimated at 3,058 million hectares (F.A.O., 1979). The rate of erosion depends on the degree of overgrazing. The estimates of soil erosion rates, expressed in kg/hectare/year, vary from less than 10 (Demek, 1969) to 25,000 (Thomas, 1956). The upper estimate would put total annual erosion from this cause at some 75,000 million tons.

Since the developing countries, between them, have three-quarters of all cattle (*U.N. Statistical Yearbook*, 1978), with the attending problems of overgrazing and poor areas, our guess as to the actual total amount involved in erosion is nearer the maximum value, taking 50,000 million tons/year.

Tilling the Land

The worldwide tilled (agricultural) area has been estimated at 1,488 million hectares (15 million km^2). The estimated rates of erosion depend on the type of crop, and range (in tons/hectare/year) from a low of 30 for wheat and barley (Demek, 1969) to a high of 180 for maize, millet, and sorghum (Bennett, 1939). With the latter constituting one-third of all crops (in terms of areas involved), our estimate as to total erosion from this cause would tend to take an average erosion of 80 tons/hectare/year, yielding for the total annual amount involved some 106,000 million tons.

The sum of the very rough estimates for the three branches of activity mentioned so far amounts to some 156,800 million tons of material eroded annually. The magnitude of the amount so moved can be illustrated by its volume, namely that of a cube 5 km end on.

Mining

Values of erosion or accumulation of material from the various mining activities are incomplete. Still, on the basis of figures of mining production (Louis, 1960; Thomas,

1956; *U.N. Statistical Yearbook*, 1978), conservative estimates put the annual amounts involved at 5,000 million tons for mining proper and at 10,000 million tons for the excavation of building materials, such as sand, gravel, and stone.

Roads and Railways

Very incomplete statistics on highway construction make even a rough estimate of the annual amounts involved in the construction of roads and railways very. difficult. The rate of erosion induced by one kilometer of a new constructed highway is estimated at 450–500 tons/year (Collier, 1972); the rate of erosion by railway tracks is negligible. It seems that the total tonnages involved annually are of the order of four hundred millions tons only.

Urban Construction

The total area under construction in 1976 has been estimated at 9,700 km^2, by extrapolating from known data for the U.S.A., (Collier, 1972) and from the U.S.A. percentage in worldwide new construction (*U.N. Statistical Yearbook*, 1977). With an annual rate of erosion of varying between 300 and 5 tons/hectare for such areas, we arrive at a total amount involved of some 17 million tons/year.

When we look once more at the admittedly tentative estimates of the total amounts of material involved in each branch of activity, we note that agricultural activities comprise the major contribution with 156,800 million tons/year, followed by mining with 15,000 (in the same units), and urban construction and means of communication with a mere 417 (idem). In the last third of the 20th century, it is still the rural activities which have the greatest anthropogeomorphological impact. Urban construction contributes no more than a fraction of some promilles.

An overall estimate of the volume of anthropogeomorphological activity results in more than 172 milliard tons/year for the middle 1970s. This would give a global per capita contribution of some 42 tons/year to the geomorphological processes. Is this figure of 172 milliard tons/year a significant one, or can it be neglected? The Huang Ho River, one of the world's muddiest rivers, carries down to the sea about 1.5 milliard tons of silt a year (*Encyclopaedia Brittanica*, 1979, Vol. 8, p. 1130). Our estimation of sediments moved by anthropic geomorphological processes is of a value more than hundred times higher. A geomorphological activity of a hundred Huang Ho's is not negligible; it is an environmental disaster.

TOWARDS A MODEL IN ANTHROPOGEOMORPHOLOGY

Introduction

Any evaluation of the future rates of anthropogeomorphologic processes encounters difficulties as a result of the unpredictability of Man's behavior. The physical laws governing the processes studied, even when well known (and such is not always the case), will represent but one side of the study, the other being the human partner. Man has to choose between a ruthless and a rational, considerate use of the environment (Fairbridge, 1971). It is mostly from a lack of knowledge of the delicacy of the environmental equilibrium (Hangen and Brown, 1971) that Man's ruthlessness stems. The present work endeavors to further such knowledge by analyzing the processes and classifying possible parameters towards evaluation of ongoing or future human intervention. It has been proposed (U.S. Dept. Interior, 1971) to employ the *rate of population growth* as a measure of the rate and magnitude of nonagricultural development and construction activity. As a parameter this factor would seem somewhat crude, since it expresses neither the economic capacity of the population and its technological competence nor the level of education and an implied perception of responsibility towards our planet's resources.

To permit first a qualitative, and then a quantitative, measure of the impact of the processes, we shall develop an "equation of potential anthropic geomorphology," which will be based on two main elements, namely the *degree of development*, representing Man's intervention, and the *degree of perception*, representing the extent of combating the erosion caused by that intervention.

The Degree of Development

The rates of anthropogeomorphological processes (AGP) are correlated with the degrees of development (DD) in Figure 62. Each "degree of development" is supposed to contain all earlier "degrees" (to its left). The rate of any AGP will be within the scale from 0 to 1; the five different degrees are spread evenly over this scale, with a controlable rate considered the average case and located accordingly.

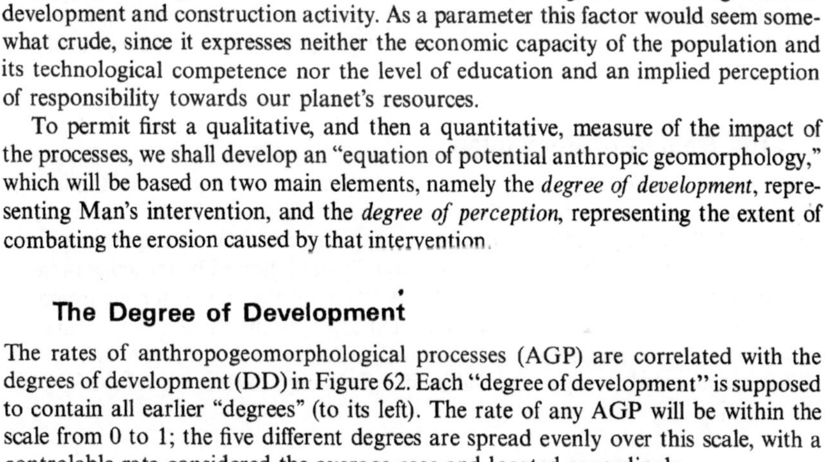

FIG. 62. Correlation between anthropogeomorphological processes and degree of Man's activity ("degree of development")

The three main possibilities of the impact of the ḊD on the rate of the AGP are represented by the three curves (Fig. 62). If the degree of development is achieved gradually, a gradually rising curve is followed, with a peak before the fullest degree of development is attained. An accelerated development process is represented by the upper curve, steep initially and descending notably further on. A delayed development will imply a curve initially horizontal, departing from this position when a certain DD will have been reached. It is assumed that all these curves end at one and the same point.

With the aid of this figure, every stage of any process (agriculture, open pit mining, building a city) can be placed relative to other processes and on the curve typifying the relevant rate of development.

The Degree of Perception

The relation between this parameter, defined above as the extent of combating the erosion, and the rate of the AGP is represented in Figure 63. Each "degree of perception" (DP) is supposed to contain all earlier, less drastic means appearing to its left in the figure. Ploughing and grazing carefully are followed by terracing of fields, and then by soil conservation and administrative protection of the environment. The possibilities of different attitudes are indicated by the three curves: a gradual development of the perception, an accelerated, and a low rate of perception. The three curves are assumed to begin and end at identical points. When the value of percep-

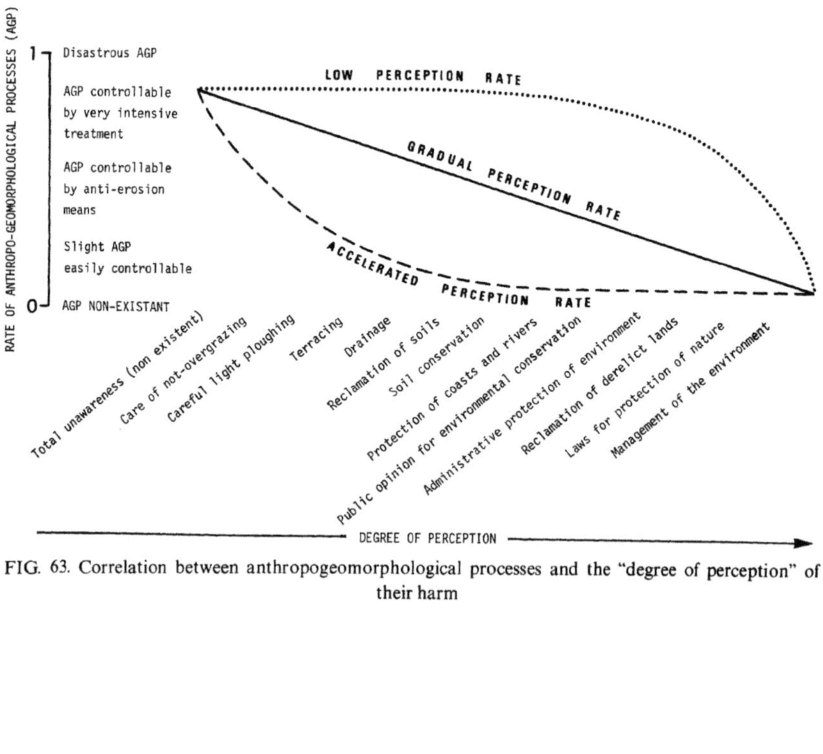

FIG. 63. Correlation between anthropogeomorphological processes and the "degree of perception" of their harm

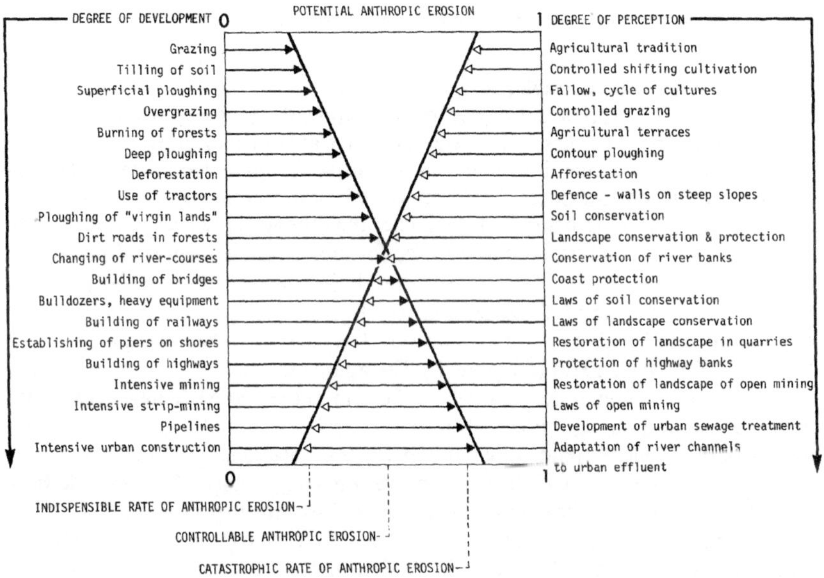

FIG. 64. Juxtaposition of degrees of development with corresponding degrees of perception

tion is low, the rate of AGP is high (when no other factors are involved); with increasing DP, the rate of the AGP decreases.

When we compare the two Figures 62 and 63, we observe that the two variables DD and DP have an opposed effect on the AGP. An increase in either influences the AGP similarly to a decrease of the other. Disregarding possible (but yet unknown, and probably unmeasurable) quantitative differences, we may put the rate of the AGP as being influenced in equal measure by an increase of either factor or a decrease of the other.

Generally speaking, three types of interaction between the DD and the DP are possible. Where the development (DD) is higher than the perception (DP), high AGP rates occur, resulting in a need for severe to moderate antierosive measures. Where both quantities are of a similar impact, there exists a certain equilibrium, and the need for corrective measures is only small. Lastly, a low AGP results from a combination of a development (DD) lower than the perception (DP).

In a semiquantitative way it is permissible to juxtapose degrees of development with corresponding degrees of perception leading to measures to combat the erosive results of such development. The changing of river courses, for example, could be connected with conservation of river banks, and intensive mining with restoration of landscape. In Figure 64 the increasing values of both degrees have been put opposite each other. The lines across the figure correspond, of course, with the "gradual rate" lines of Figures 40 and 41.

A Quantitative Measure of Man's Influence on the AGP

It is tempting to try and replace the qualitative notions of development (DD) and perception (DP) by measurable, quantitative entities.

We propose to express the degree of development by the *percentage of urban population*, taking, of course, into account that statistics indicating merely size of a settlement without mentioning the inhabitants' occupation may be misleading. Further, we suggest to express the degree of lack of perception by the *percentage of illiteracy*, the reasoning being that the higher the illiteracy, the lower the education, education being a necessary condition for forming a public degree of perception.

To check on the validity of the implied correlation between the two chosen values, the data (where available) of the percentage of illiteracy versus the percentage of urban population have been set out for 37 countries (*U.N. Statistical Yearbook*, 1977; *Unesco Statistical Yearbook*, 1977) in Figure 65. The figure indicates a negative correlation, with most values lying in the band where the sum of the two percentages is approximately constant. Four countries — Iraq, Morocco, Tunisia, and Algeria — seem to lie away from the general trend, apparently because of insufficiency of the definition of "urban population" when applied to a population living in oases. On the whole, however, the figure sustains the correctness of our choice of the two variables.

Denoting now the percentage of urban population by UP, and the percentage of illiteracy as DI, we define the rate of the AGP as the average value of these two variables, i.e., (UP + DI)/2. This rate will preferably be expressed in percentages. If we set this rate out against the UP (see Fig. 66), it will be noted that most countries lie at or below a rate of AGP of 50%, which we consider still relatively safe. India, for

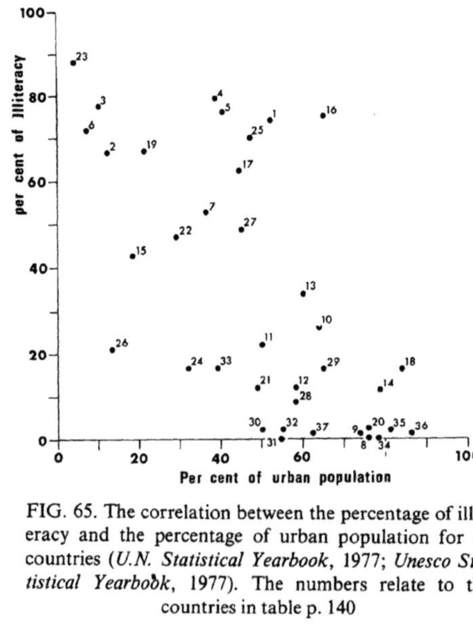

FIG. 65. The correlation between the percentage of illiteracy and the percentage of urban population for 37 countries (*U.N. Statistical Yearbook*, 1977; *Unesco Statistical Yearbook*, 1977). The numbers relate to the countries in table p. 140

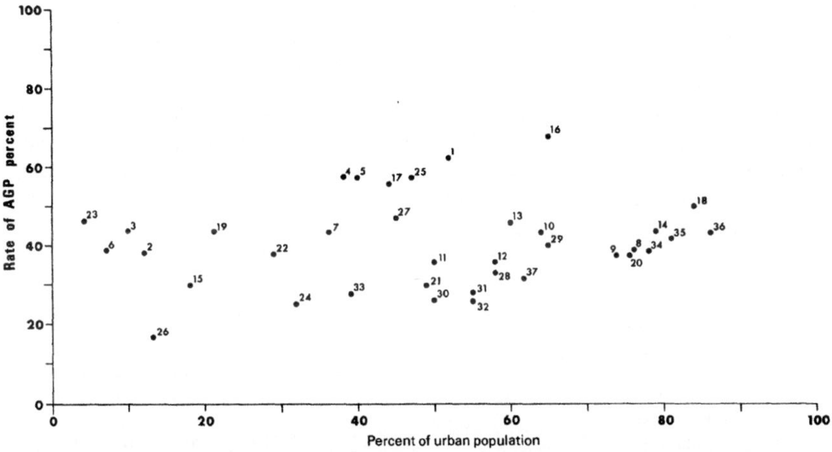

FIG. 66. The rate of anthropogeomorphological processes (AGP) defined as the average of percentage of urban population (UP) and percentage of illiteracy (DI), plotted against the UP

example, showing a higher rate than the U.S.A., should take more stringent measures than the latter. Six countries (Algeria, Morocco, Tunisia, Zambia, Iran, and Iraq) with a really high rate of AGP require urgent and strong measures.

Integration of the Human and the Environmental Factors

The rate of AGP, as expressed above, represents Man's part only in these processes. To include all physical factors in one general equation with the human ones would seem a fairly hazardous task. We propose, however, to integrate into a general equation the two most important physical factors, namely those of *climate* and of *relief.* We define an INDEX OF POTENTIAL ANTHROPIC GEOMORPHOLOGY, *I*, in the following way:

$$I = \frac{\text{UP} + \text{DI}}{2} \times \frac{1}{100} \times (K_c + K_r)$$

where UP and DI retain their former meaning; the multiplication by 1/100 serves to express the average of UP and DL by a value between 0 and 1; K_c = a constant due to climatic conditions; and K_r = a constant due to relief conditions. For the two constants, the following numerical values are proposed:

The value of K_c will be as set out against the proper climate (classification after Köppen):

Equatorial climate	Ae	0.6
Monsoon-savannah climate	Aw	0.8
Arid and semiarid climate	B	0.6
Temperate climate	C	0.4
Cold climate	D	0.6
Arctic climate	E	0.4

The value of K_r will be set out against the proper relief:

Plains	0.2	Medium-high mountains	0.6
Hills	0.4	High (Alpine) mountains	0.8
Plateaus	0.5		

Whereas the relative values of these constants have been realized, their absolute numerical values are proposed here for the first time. The index obtained by multiplication of the two factors will usually have a value between 0 and 1. The following table contains values of I for 37 countries.

Values of index of potential anthropic geomorphology*

No.	Country	Urban population (UP) (%)	Illiteracy (DI) (%)	$\frac{(UP+DI)/2}{100}$	K_c	K_r	I
1	Algeria	52	74	0.63	0.6	0.6	0.76
2	Botswana	12	67	0.39	0.6	0.2	0.31
3	Malawi	10	78	0.44	0.7	0.4	0.48
4	Morocco	38	79	0.58	0.6	0.6	0.70
5	Tunisia	40	76	0.58	0.6	0.4	0.58
6	Tanzania	7	72	0.39	0.8	0.5	0.50
7	Zambia	36	53	0.44	0.8	0.5	0.57
8	Canada	76	1.5	0.39	0.6	0.5	0.48
9	U.S.A.	74	1	0.38	0.5	0.4	0.34
10	Mexico	64	26	0.44	0.6	0.5	0.48
11	Panama	50	22	0.36	0.8	0.4	0.43
12	Puerto Rico	58	12	0.35	0.8	0.4	0.42
13	Brazil	60	34	0.46	0.7	0.3	0.46
14	Chile	79	12	0.44	0.6	0.5	0.48
15	Indonesia	18	43	0.30	0.8	0.4	0.36
16	Iraq	65	75	0.68	0.6	0.3	0.46
17	Iran	44	63	0.55	0.6	0.5	0.60
18	Israel	84	16	0.50	0.6	0.4	0.50
19	India	21	67	0.44	0.8	0.5	0.57
20	Japan	76	2	0.39	0.6	0.4	0.39
21	Korea	49	12	0.30	0.5	0.4	0.27
22	Malaysia	29	47	0.38	0.8	0.4	0.36
23	Nepal	4	88	0.46	0.8	0.7	0.70
24	Philippines	32	17	0.25	0.8	0.5	0.33
25	Syria	47	70	0.58	0.6	0.3	0.52
26	Thailand	13	21	0.17	0.8	0.2	0.17
27	Turkey	45	49	0.47	0.5	0.5	0.47
28	Bulgaria	58	9	0.33	0.4	0.6	0.33
29	Greece	65	16	0.40	0.4	0.6	0.40
30	Hungary	50	2	0.26	0.4	0.2	0.16
31	Poland	55	2	0.28	0.4	0.2	0.17
32	Switzerland	55	0.2	0.28	0.4	0.8	0.34
33	Yugoslavia	39	17	0.28	0.4	0.4	0.22
34	United Kingdom	78	1	0.39	0.3	0.4	0.35
35	New Zealand	81	2	0.42	0.5	0.5	0.42
36	Australia	86	1	0.44	0.5	0.3	0.35
37	U.S.S.R.	62	0.3	0.31	0.5	0.4	0.28

* Data from the *U.N. Statistical Yearbook*, 1977; *Unesco Statistical Yearbook*, 1977

The distribution of the values of I among the various countries is given in the following table:

Range of I	Number of countries	% of total
0.10–0.19	3	8
0.20–0.29	3	8
0.30–0.39	10	27.3
0.40–0.49	11	30
0.50–0.59	6	16
0.60–0.69	1	2.7
0.70–0.79	3	8

The distinct grouping is represented graphically in Figure 67. We note three main categories of countries:

1. Six countries (16%) with a very low index I (<0.29), comprising Korea, Thailand, Hungary, Poland, Yugoslavia, and the U.S.S.R. These countries depict a small danger only of Man's adverse geomorphological actions.

2. Twenty-one countries (57%) with a medium index, still below the critical value of 0.50 ($0.30 < I < 0.49$), comprising Botswana, Malawi, Canada, the U.S.A., Mexico, Panama, Puerto Rico, Brazil, Chile, Indonesia, Iraq, Japan, Malaysia, the Philippines, Turkey, Bulgaria, Greece, Switzerland, the United Kingdom, New Zealand, and Australia, where the mentioned danger is not negligible and measures to combat the threats are required.

3. Ten countries (over 26%) of a high index I ($0.50 < I < 0.79$), comprising Algeria, Morocco, Tunisia, Tanzania, Zambia, Iran, Israel, India, Nepal, and Syria, where the risks posed by Man on the environment are very great, and, unless these risks are energetically and thoroughly tackled, they will result in considerable harm.

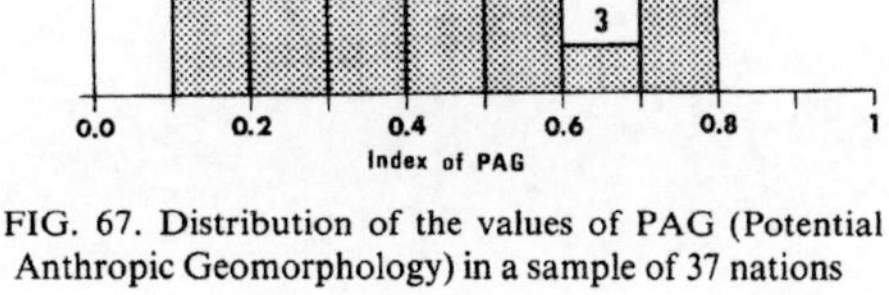

FIG. 67. Distribution of the values of PAG (Potential Anthropic Geomorphology) in a sample of 37 nations

Appendix: The Influence of Warfare

INDIRECT INFLUENCE OF WARFARE

The indirect influence of warfare is represented by the damage to normal land use, as in the destruction of irrigation canals, the obstruction of rivers, and so on. Vita-Finzi (1963), examining the Mediterranean streams of North Africa, discovered alluvial deposits some three to eight meters deep, which were formed after the stone dams, built in Roman times, had collapsed. Guy (1954) described a similar phenomenon, namely, widespread soil erosion in different parts of North Africa and the Levant after the Classic period. Vita-Finzi attributes this erosion to the political disturbances which followed the decline of the Roman Empire, which were, at least partly, accompanied by warfare. There are certainly other areas where similar effects can be noticed, but in the Mediterranean and other arid climates this influence is perhaps more conspicuous. Another indirect effect of warfare were the dust storms generated by the movements of tanks and other vehicles in the North African campaigns in 1941–1943. Also, it is known that in the three years following the end of World War II, when the banks of the Rochefort basin of France were still destroyed (as a result of the military operations of 1944), seven meters of sediment accumulated in the ports of the Charente estuary (Vita-Finzi, 1974).

DIRECT INFLUENCE OF WARFARE

The extent of the direct influence of warfare on the landscape grew with the development of war technology; it was only with the introduction of trenches and bombing in World War I that significant changes in the relief occurred as a result of direct warfare. Smaller changes were, however, induced already in earlier periods. In 634 A.D., the Byzantine town of Skythopolis (Beth She'an) in the Jordan Valley opened the irrigation ditches to inundate the fields and thereby to slow down the advance of the Muslim conquerors; the effect was a temporary return of the land to swamps, as it had been before the management of the springs of the valley (Nir, 1968). The same use of ditches, dikes, and water diversions as an instrument of warfare has been reported from China since the 5th century B.C. (Klimm, 1956).

More consequential use of ditches for warfare, and, as a result, a change of the terra firma into marshes was made in The Netherlands (Wagret, 1958): in 1672, a deliberately caused flooding barred the way to the conquering soldiers of Louis XIV. In 1942, at the time of the Allied landings in North Africa, the Germans, fearing an invasion, released vast stretches of water to cut off the middle of Holland; in 1944, this was repeated in the Meuse and Rhine regions. In 1944/45, the Dutch island of Walcheren was nearly wiped off the map, when the Allies decided to flood the whole island. Dikes were bombed, and that of Westkapelle, the strongest in Europe and 100 meters wide, disappeared beneath the water. The total length of breaches induced by this interference was 2,200 meters. During the war years 1940–1945, some 230,000 hectares were affected in all by flooding, induced by warfare tactics.

As modern warfare employs more and more technologically "advancing" means and constructions of roads, airfields, etc., interferences with the landscape occur which would not otherwise have happened. In Vietnam, military construction activities in the 1960s exposed large areas of soil and subsequently led to serious erosion problems (Hay et al., 1970). The monsoon rains and winds easily eroded the man-made steep slopes of revetments and other structures demanded by military operations.

The actual battlefields demonstrate, however, the harshest influence of warring. They have become a barren, uncultivable area of trenches, craters, and shattered terrain. One of the highest elements in the relief of Berlin today consists of ten mounds, the biggest of which is 115 meters high, built from material removed from buildings destroyed during World War II. Their total volume is nearly 100 million m^3 (Vita-Finzi, 1974).

Bibliography

Adamson, C. M. (1974): Effects of soil conservation treatment on runoff and sediment loss from a catchment in southwestern NSW, Australia, *IAHS-AISH Spec. Publ.* **113**, 3–14.

Aéroport de Paris, Charles de Gaulle (1974): Editions Publicis, Paris.

Aghassy, J. (1973): Man-induced badlands topography. In D. R. Coates (ed.), *Environmental Geomorphology and Landscape Conservation. III, Non-Urban,* Benchmark Papers in Geology **8**, 124–136.

Anderson, D. G. (1970): Effects of urban development on floods in northern Virgina, *U.S. Geol. Surv. Wat. Serv. Pap.* 2001-C.

Anderson, H. W. (1974): Sediment deposition in reservoirs associated with rural roads, forest fires and catchment attributes, *IAHS-AISH Spec. Publ.* **113**, 87–95.

Arber, M. (1946): Dust storms in the Fenlands round Ely, *Geogr.* **31**, 23–26.

Arvill, R. (1969): *Man and Environment*, Penguin, London.

Aufrère, L. (1929): Les rideaux—étude topographique, *Annls Géogr.* **38**, 529–560.

Bailey, R. G. (1973): Forest use implication. In D. R. Coates (ed.), *Environmental Geomorphology and Landscape Conservation, III, Non-Urban,* Benchmark Papers in Geology **8**, 388–413.

Ball, D. F. (1975): Processes of soil degradation: a pedological point of view. In J. J. Evans, S. Limbrey and H. Cleere (eds.), *The Effect of Man on the Landscape: the Highland Zone*, Counc. for Brit. Archaeolog., Res. Pap. No. 11, pp. 20–27.

Barr, D. A. (1965): Restoration of coastal dunes after beach mining, *Journ. Soil Conserv. Serv., N.S. Wales* **21**, 199–209.

Barr. D. A. and W. J. Atkinson (1970): Stabilisation of coastal sands after mining, *Journ. Soil Conserv. Serv., N.S. Wales* **26**, 89–107.

Barr, J. (1970): *Derelict Britain*, Pelican, London.

Beatty, R. A. (1966): The inert becomes "ert", *Landscape Architecture* 125–128.

Beaumont, P. (1978): Man's impact on river systems, a world-wide view, *Area* **10**, 38–42.

Becker, J. C. (1972): Alaska builds highway over muskeg and permafrost, *Civil Eng., ASCE* **42**, 75–77.

Bennett, H. H. (1928): Soil erosion—a national menace, *U.S. Dept. of Agr. Circular* **33** (repr. in *Benchmark Papers in Geology* **8**, 57–83).

Bennett; H. H. (1939): *Soil Conservation*, New York.

Bennett, H. H. (1955): *Elements of Soil Conservation*, 2nd edn., New York.

Bidwell, O. W. and F. D. Hole (1965): Man as a factor of soil formation, *Soil Sci.* **99**, 65–72.

Birot, P. (1966): *General Physical Geography*, London.

Bishop, D. M. and M. E. Stevens (1964): Landslides on logged areas in South-East Alaska, North. Forest Exp. Stat., Juneau, *U.S. Forest Serv. Res. Pap.*, NOR-1.

Black, P. E. (1971): Interrelationship of forest, soils and terrain in watershed planning. In D. R. Coates (ed.), *Environmental Geomorphology*, Binghampton, pp. 71–78.

Bowmaker, A. (1970): A prospect of Lake Kariba, *Optima* 68–74.
Božovič, A. (1974): Review and appraisal of case histories related to seismic effects of reservoir impounding, *Eng. Geol.* **8**, 9–27.
Brater, E. F. and S. Sangal (1969): Effects of urbanisation on peak floods. In W. D. Moore and C. W. Morgan (eds.), *Effects of Watershed Changes on Streamflow*, Univ. of Texas, Austin, pp. 201–214.
Brown, E. H. (1970): Man shapes the earth, *Geogrl. Journ.* **136**, 74–85.
Bull, W. B. and R. E. Miller (1975): Land subsidence due to ground water withdrawal in the Los Banos-Kettleman City area, Cal., Part I, Changes in the hydrologic environmental conduction to subsidence, *U.S. Geol. Surv. Prof. Pap.* 437-E.
Bullock, W. D., R. L. Brittain and G. A. Place (1975): Environmental aspects of the NA mining industry. In M. M. Jones (ed.), *Minerals and Environment*, Inst. of Min. and Metallur., pp. 27–43.
Cailleux, A. and L. E. Hamelin (1969): Poste de Baléine (Nouveau-Québec): Exemple de géomorphologie complexe, *R.d.G.D.* **3**, 129–150.
Cantor, L. M. (1967): *A World Geography of Irrigation*, London.
Carder, D. C. (1968): Reservoir loading and local earthquakes, *Ann. Eng. Geol. Soils Eng. Symp.*, Boise, Idaho.
Caris, S. and M. Pavish (1975): *Bibliography: Environmental Geomorphology*, Council of Planning Librarians, P.O.B. 229, Monticello, Ill.
Carter, A. W. (1961): Magnitude and frequency of floods in suburban areas, *U.S. Geol. Surv. Prof. Pap.* **424-B**, 9–11.
Chandler, T. and C. Fox (1974): *3000 Years of Urban Growth*, Academic Press, New York.
Chen, Ch. N. (1974): Effects of land development on soil erosion and sediment concentration in an urbanising basin, *IAHS-AISH Spec. Publ.* **113**, 150–157.
Chi-sen Liang (1973): *Urban Land-Use Analysis; A Case Study of Hong-Kong*, London.
Chorley, R. I. (1962): Geomorphology and general system theory, *U.S. Geol. Sur. Prof. Pap.* **500**.
Clark, A. H. (1949): *The Invasion of New Zealand by People, Plants and Animals*, Rutgers, New Brunswick.
Coates, D. R. (ed.) (1971): *Environmental Geomorphology*, State University of New York at Binghamton, Publ. in Geomorphology.
Coates, D. R. (ed.) (1973): *Coastal Geomorphology*, State University of New York at Binghamton, Publ. in Geomorphology.
Coates, D. R. (ed.) (1973a): *Environmental Geomorphology and Landscape Conservation, III: Non-Urban*, Benchmark Papers in Geology, **8**.
Coates D. R. (ed.) (1976): Geomorphology and engineering, *Proc. 7th Amer. Geomorph. Symposium*, State University of New York at Binghamton 24–25.9.1976.
Collier, C. R. (ed.) (1972): Committee on Geological Sciences: *The Earth and Human Affairs*, Nat. Acad. of Sci., U.S.A., Division of Earth Sci., San Francisco.
Collier, C. R., R. J. Pickering and J. J. Musser (1970): Influence of strip mining on the hydrologic environment of parts of Beaver Creek basin, Kentucky, 1955/56, *U.S. Geol. Surv. Prof. Pap.* 427-C.
Conacher, A. J. (1971): The significance of vegetation, fire and man in the stabilisation of sand dunes near Warburton Ranges, C. Australia, *Earth Sci. Journ. (New Zealand)* **5**, 92–94.
Conacher, A. J. (1974): Salt scalds, *Austr. Sci. and Technol.* **11**, 14–16.
Conacher, A. J. and I. D. Murray (1973): Implications and causes of salinity problems in the Western Australia wheatbelt: the York-Mawson area, *Austr. Geogrl. Stud.* **11**, 40–61.
Condliffe, J. B. (1936): *New Zealand in the Making*, London.
Cooke, R. U. and J. C. Doornkamp (1974): *Geomorphology in Environmental Management*, Oxford University Press, London.
Cronin, L. E. (1971): The role of man in estuarine processes. In T. R. Detwyler (ed.), *Man's Impact on Environment*, New York, pp. 266–294.
Curry, R. R. (1971): Soil destruction associated with forest management and prospects for recovery in geological time, *Ass. Southeastern Biol. Bull.* **18**, 117–128.
Curtis, J. T. (1956): The modification of mid-latitude grasslands and forests by man. In W. L. Thomas (ed.), *Man's Role in the Changing the Face of the Earth*, Univ. of Chicago Press, Chicago, pp. 721–735.
Curtis, W. B. (1971): Terraces reduce run-off and erosion on surface-mine benches, *Journ. of Soil and Water Conserv.* **26**, 198–199.

Daniels, R. B. (1960): Entrenchment of the Willow River drainage ditch, Harrison Country, Iowa, *Amer. Journ. Sci.* **258**, 161–176.

Danserau, P. (ed.) (1970): *Challenge for Survival: Land, Air and Water for Man in Megalopolis*, New York.

Danserau, P. (1971): Ecologie de la zone de l'aéroport international de Montréal. Une aventure interdisciplinaire, *Rev. Géogr. Montréal* **26**, 301–305.

Darby, H. C. (1956): The clearing of the woodland in Europe. In W. L. Thomas (ed.), *Man's Role in Changing the Face of the Earth*, Univ. of Chicago Press, Chicago, pp. 183–216.

Davies, I. V., C. Brook and R. T. Arguile (1975): Opencast coal mining; working, restauration and reclamation. In M. S. Jones (ed.), *Minerals and Environment*, Inst. of Min. and Metallur., London, pp. 313–331.

Davis, J. H. (1956): Influences of man upon coast lines. In W. L. Thomas (ed.), *Man's Role in Changing the Face of the Earth*, Univ. of Chicago Press, Chicago, pp. 504–521.

Davis, W. M. (1899): The geographical cycle, *Geogrl. Journ.* **14**, 481–504.

Demangeon, A. (1927): Belgique-Pays-Bas-Luxembourg. In P. Vidal de la Blache and L. Galais (eds.), *Géographie Universelle*, Tome II, pp. 25–45.

Demek, J. (1969): Beschleunigung der geomorphologischen Prozesse durch die Wirkung des Menschen, *Geol. Rundschau* **58**, 111–121.

Denevan, W. M. (1970): Aborginal drained-field cultivation in the Americas, *Science* **169**, 647–654.

Dept. of Environment (1972): *Maplin Airport—Choice of sites for runways.*

Derruau, M. (1967): *Précis de géomorphologie*, 5th edn., Masson, Paris.

Detwyler, T. R. (ed.) (1971): *Man's Impact on Environment*, New York.

Diseker, E. G. and E. C. Richardson (1961): Roadside sediment production and control, *Trans. Amer. Soc. Agr. Eng.* 62–68.

Diseker, E. G. and E. C. Richardson (1962): Erosion rates and control methods in highway cuts, *Trans. Amer. Soc. Agr. Eng.* 153–155.

Doerr, A. and L. Guernsey (1956): Man as a geomorphological agent; the example of coal mining. *A.A.A.G.* **46**, 197–210.

Dolan, R. (1973): Barrier islands: natural and controlled. In D. R. Coates (ed.), *Coastal Geomorphology* State University of New York at Binghamton, pp. 263–278.

Douglas, I. (1967a): Natural and man-made erosion in the humid tropics of Australia, Malaysia and Singapore, *IAHS-AISH Spec. Publ.* **75**, 17–30.

Douglas, I. (1967b): Man, vegetation and the sediment yield of rivers, *Nature* **215**, 925–928.

Douglas, I. (1976): Urban hydrology, *Geogrl. Journ.* **142**, 65–72.

Dresch, J. (1977): The evaluation and exploitation of the West African Sahel, *Philosoph. Transact. Roy. Soc., Series B* **278**, No. 962, 537–542.

Duchaufour, P. (1965): *Précis de Pédologie*, 2nd edn., Masson, Paris.

Dury, G. H. (1974): Rivers and river systems, *Encycl. Brit.*, 1974 edition.

Eakin, H. M. (1936): Silting of reservoirs. In D. R. Coates (ed.), *Environmental Geomorphology and Landscape Conservation*, Benchmark Papers in Geology **8**, 87–102.

Eckholm, E. P. (1976): *Losing Ground*, New York.

El-Ashry, M. T. (1971): Causes of recent increased erosion along U.S. shoreline, *Geol. Soc. Amer. Bull.* **82**, 2033–2038.

Eliassen, S. (1936): Soil erosion and river regulation with special reference to the Yellow River. In D. R. Coates, (ed.), *Environmental Geomorphology and Landscape Conservation*, Benchmark Papers in Geology **8**, 325–344.

Evans, J. G., S. Limbrey, and H. Cleere, (eds.) (1975): The effect of man on the landscape: the Highland zone, Counc. for Brit. Archaeology *Res. Pap. no. 11.*

Evenari, M., L. Shanan and N. H. Tadmor (1963): Run-off farming in the Negev Desert of Israel, report on the Ovdat and Shivta projects for the years 1958–1962, *Spec. Publ. Nat. Univ. Inst. Agric.* Rehovot (Israel), No. 393-A.

Eyles, R. S. (1971): Mass movement in Tangoio conservation reserve, Northern Hankes bay, *Earth Sci. Journ. (New Zealand)* **5**, 79–81.

Fairbridge, R. W. (1971): Society and geomorphology. In D. R. Coates, (ed.), *Environmental Geomorphology*, State University of New York at Binghamton, pp. 215–220.

Fairbrother, N. (1972): *New Lives, New Landscapes*, Pelican, London.

FAO (1971): *Report on the 1960 World Census of Agriculture*, Rome.
FAO (1972): *The State of Food and Agriculture, 1972*, Rome.
FAO (1977): *FAO Production Yearbook*, vol. 31.
Fels, E. (1935): *Der Mensch als Gestalter der Erde*, Leipzig.
Fischer, E. (1915): Der Mensch als geologischer Faktor, *Zeitsch. Deutsch. Geolog. Geselsch.* **67**, 106–148.
Flawn, P. T. (1970): *Environmental Geology*, New York.
Floods in Eastern Sinai. No. 3. (1972): *The floods of 23–25.11.1972, and their relation to the Eilat-Ophira highway*, Hebr. Univ. Jerusalem Dept. of Geography (Hebrew, polycopied).
Fournier, F. (1960): *Climat et erosion*, P.H.F., Paris.
Fournier, F. (1972): Utilisation rationelle et conservation du soil, *Geoforum* **10**, 35–47
Game, H. F. (1974): *Origins of Landscapes*, New York.
Gautier, M. (1975): Actions naturelles et actions humaines dans l'évolution d'une flèche littorale: le cordon dunaire des Moutiers en Retz (L.M.), *Norois* **22**, 549–562.
Gillette, R. (1972): Stream channelization: conflict between ditchers and conservationalists, *Science* **176**, 890–894.
Glenn, L. Ch. (1911): Denudation and erosion in the southern Appalachian region and the Monongahela Basin, *U.S. Geol. Surv. Prof. Pap.* **72**, 10–30.
Godfrey, R. J. and M. M. Godfrey (1973): Comparison of ecological and geomorphic interactions between altered and unaltered Barrier Island systems in North Carolina. In D. R. Coates, (ed.), *Coastal Geomorphology*, State University of New York at Binghamton, pp. 239–258.
Golomb, B. and H. M. Eder (1964): Landforms made by man, *Landscape* 4–7.
Gottmann, J. (1957): *Les manchés des matières prémières*, Paris.
Gourou, P. (1955): Une expérience d'agriculture mécanisée en Afrique Orientale: le Plan des Arachides, *C.O.M.* **8**, 105–118.
Gourou, P. (1966): *The Tropical World*, 2nd edn., London.
Gov. of South Africa, *Mining Survey*, 1975.
Graf, L. (1976): Stream, slopes and suburban development, *Geogrl. Analysis* **8**, 153–173.
Graham, E. H. (1956): The recreative power of plant communities. In W. L. Thomas (ed.), *Man's Role in Changing the Face of the Earth*, Univ. of Chicago Press, Chicago, pp. 677–691.
Gray, D. H. (1969): Effects of Forest Clear Cutting on the Stability of Natural Slopes, Univ. of Michigan, Dept. of Civil Engineering, Ann Arbor, Nat. Sci. Found. Grant No. GK-2377.
Greer, J. D. (1971): Effects of excessive-rate rainstorms on erosion, *Journ. of Soil and Water Conserv.* **26**, 196–197.
Gregory, K. S. and D. E. Walling (1980): *Man and Environmental Processes*, Butterworths.
Guilcher, A. (1965): *Précis d'hydrologie marine et continentale*, Masson, Paris.
Guy, H. P. (1970): Sediment problems in urban areas, *U.S. Geol. Surv. Circular* **601-E**, E1–E8.
Guy, H. P. (1974): Remote sensing techniques for evaluation of urban erosion and sedimentation, *IAHS-AISH Spec. Publ.* **113**, 145–149.
Guy, H. P. and G. E. Ferguson (1970): Stream sediment: an environmental problem, *Journ. of Soil and Water Conserv.* **25**, 217–221.
Guy, P. L. O. (1954): Archaeological evidence of soil erosion and sedimentation in Wadi Musrara, *IEJ* **4**, 77–87.
Hadley, R. F. (1974): Sediment yield and land use in southwest U.S., *IAHS-AISH Spec. Publ.* **113**, 96–98.
Hadley, R. F. (1977): Evaluation of land-use and land-treatment practices in semi-arid western United States, *Philosoph. Transact. Roy. Soc., Series B* **278**, 543–554.
Haigh, M. J. (1978): *Evolution of slopes on artificial landforms, Blainarch. U.K.*, Univ. of Chicago, Dept. of Geol. Res. Papers No. 183.
Hamilton, D. H. and R. L. Meehan (1971): Ground rupture in the Baldwin Hills, *Science* **172**, 333–344.
Hangen, R. K. and J. Brown (1971): Natural and man-induced disturbances of permafrost terrain. In D. R. Coates, (ed.), *Environmental Geomorphology*, State University of New York at Binghamton, pp. 139–149.
Hansen, E. (ed.) (1975): Inadvertent effects of man on the hydrological cycle, *Nordic Intern. Hydrolog. Decade Report No. 8*, Nat. Comm. for the Intern. Hydr. Decade in Denmark, Finland, Iceland, Norway and Sweden.

Hansen, E. (1975): Groundwater development and its influence on the runoff characteristics. In Hansen, E. (ed.), *Nordic Intern. Hydrolog. Decade Report No. 8*, pp. 50–53.

Harris, E. E. and S. E. Rantz (1964): Effect of urban growth on streamflow regimen of Permanence Creek, Santa Clara Country, Cal., *U.S. Geol. Surv. Water Supply Paper* 1591-B.

Hartwright, T. U. (1975): Development of gravel-pit lakes for leisure purposes. In M. J. Jones (ed.), *Minerals and the Environment*, Inst. of Min. and Metallur., pp. 333–339.

Haserodt, K. (1971): Reliefveränderungen durch Grossterrassen in der Lösslandschaften des südlichen Oberrheingebiete, — Ein Beitrag zur anthropogenen Geomorphologie, *Regio Basiliensis* **12**, 330–354.

Hay, D. R., D. F. Kibler and C. E. Busby (1970): Erosion control on Air Force bases, Ann. Meet. Amer. Soc. Agr. Eng., August 1940, *Technical Note WLC-TH-80-018, Civil Eng. Div. Air Forc. Weap./Lab.*, Kirtland Air Force Base.

Hayes, C. J. (1971): Landslides and related phenomena pertaining to highway construction in Oklahoma. In *Environmental Aspects of Geology and Engineering*, Symp., Annals Okla. Acad. of Sci. Publ., Vol. 2, pp. 47–57.

Hempel, L. (1971): Die Tendenzen anthropogen bedingten Reliefformung in den Ackerlaendern Europas, *ZfG* **15**, 312–329.

Hilpman, P. L. (1970): Urban growth and environmental geology, *Colorado Geol. Surv. Spec. Publ.* **1**, 16–19.

Holz, R. K. (1969): Man-made landforms in the Nile delta, *Geogrl. R.* **59**, 251–269.

Hornbeck, J. W., R. S. Pierce and C. A. Federer (1970): Streamflow changes after forest clearing in New England, *Water Res. Res.* **6**, 1124–1132.

Horton, R. E. (1945): Erosional development of streams and their drainage basins; hydrophysical application to quantitative morphology, *Geol. Soc. Amer. Bull.*, **56**, 275–310.

Howard, A. D. and I. Remson (1978): *Geology in Environmental Planning*, McGraw Hill.

Humbert, A. (1975): Les ruptures de pente dans les terres cultivées: Etude de phénomène en Andalusie, *R. d. g. de l'Est* **15**, 245–271.

Jacks, G. V. and R. O. Whyte (1939): *The Rape of the Earth*, London.

Jacobsen, T. and R. M. Adams (1958): Salt and silt in ancient Mesopotamian agriculture, *Science* **128**, 1251–1258.

Jennings, J. N. (1965): Man as a geological agent, *Austr. Journ. of Sci.* **28**, 150–156.

Jepson, H. G. (1939): Prevention and control of gullies, *U.S. Dept. of Agric., Farmers' Bull.* **1813**, 1–33.

Johnson, A. W. (1961): Highway erosion control, *Transac. Amer. Soc. Civ. Eng.* 144–152.

Jones, D. K. C. (1973): Man moulds the landscape, *Geogrl. Mag.* **65**, 575–581.

Judd, W. R. (1974): Introduction (to the International Colloquium on Seismic Effects of Reservoir Impounding), *Eng. Geol.* **8**, 1–3.

Judson, S. (1963): Erosion and deposition of Italian stream valleys during historic times, *Science* **140**, 898–899.

Judson, S. (1968): Erosion rates near Rome, Italy, *Science* **160**, 1444–1445.

Karmon, Y. (1960): The drainage of the Hula swamps, *Geogrl. R.* **50**, 169–193.

Kedar, Y. (1957a): Ancient agriculture at Shivtah in the Negev, *I.E.J.* **7**, 178–189.

Kedar, Y. (1957b): Water and soil from the desert: some ancient agricultural achievements in the Central Negev, *Geogrl. Journ.* **123**, 178–187.

Kedar, Y. and Y. Ziv (1964): The supply of water in Caesarea, *Bull. Isr. Expl. Soc.* **28**, 122–131 (in Hebrew).

Keesing's Contemporary Archives, 1965/66, 21683A, London.

Keesing's Contemporary Archives, 1967/68, 27289, London.

Keller, E. A. (1976): Channelisation—environmental, geomorphic and engineering aspects. In D. R. Coates (ed.), *Geomorphology and Engineering*, University of New York at Binghamton, pp. 115–140.

Keller, F. J. (1962): Effect of urban growth on sediment discharge, NW. branch, Anacostia River Basin, Maryland, *U.S. Geol. Surv. Prof. Pap.* 450-C.

King, H. J., B. N. Whittaker and C. H. Shadbolt (1975): Effects of mining subsidence on surface structures. In M. S. Jones (ed.), *Minerals and the Environment*, Inst. of Min. and Metallur., pp. 617–642.

Kinosita, T. and Y. Yamazaki (1974): Increase of sediment transport due to large-scale urbanisation, *IAHS-AISH Spec. Publ.* **113**, 127–129.

Klimm, L. E. (1956): Man's ports and channels. In W. L. Thomas (ed.), *Man's Role in Changing the Face of the Earth*, Univ. of Chicago Press, Chicago, pp. 522–541.

Kollmorgen, W. M. (1953): Settlement control beats flood control, *Econ. Geogr.* **29**, 208–215.

Kotlov, F. V. (1972): Modification in the geological environment resulting from growth of cities, Geolog. Surv. of Canada, 1971, *24th Intern. Congr. Geol:, Montreal, Symposium I*, Eng. Geol., Ottawa, pp. 36–40.

Lal, B. V. and S. Baneyi (1974): Man's impact on erosion in the rural environment: the Indian experience, *IAHS-AISH Spec. Publ.* **113**, 46–52.

Legget, R. F. (1962): *Geology and Engineering*, 2nd edn., New York.

Legget, R. F. (1973): *Cities and Geology*, New York.

Leighton, F. B. (1972): Origin and control of landslides in the urban environment of California, Geolog. Survey of Canada, 1972, *24th Intern. Congr. Geol., Montreal, Section 13*, Eng. Geol., Ottawa, pp. 89–96.

Leopold, L. B. (1956): Land use and sediment yield. In W. L. Thomas (ed.), *Man's Role in Changing the Surface of the Earth*, Univ. of Chicago Press, Chicago, pp. 639–647.

Leopold, L. B. (1968): Hydrology for urban land planning—a guidebook on the hydrologic effects of urban land-use, *U.S. Geol. Surv. Circular* No. 554.

Leopold, L. B., J. P. Miller and M. G. Wolman (1964): *Fluvial Processes in Geomorphology*, San Francisco.

Lewin, J. (1975): Geomorphology and environmental impact statements, *Area* **7**, 2, 127–129.

Lindh, G. (1972): Urbanisation, a hydrological headache, *Ambio* **1**, 185–201.

Louis, H. (1960): *Allgemeine Geomorphologie*, Berlin.

Lowdermilk, W. C. (1927): Erosion control in Japan, In D. R. Coates (ed.), *Environmental Geomorphology and Landscape Conservation*, Benchmark Papers **8** (1973), 316–324.

Lull, H. W. (1971): Runoff from forest lands. In W. H. Mathews, F. L. Smith and E. D. Goldberg (eds.), *Man's Impact on Terrestrial and Oceanic Ecosystems*, MIT Press, Cambridge, pp. 240–251.

Maderey, L. E. (1974): Effect de la croissance de la partie ouest de la ville de Mexico sur les processus hydrodynamiques, *IAHS-AISH Spec. Publ.* **113**, 130–136.

Makhaveyev, N. I. (1972): The impact of large water engineering projects on geomorphic processes in stream valleys, *Soviet Geogr.* **13**, 387–393.

Manners, I. R. and M. W. Mikessell (1974): *Perspectives on Environment*, A.A.A.G., Washington.

Marsh, G. P. (1864): *Man and Nature, or Physical Geography as Modified by Human Action* (edited 1965 by D. Lowenthal), New York.

Martens, L. A. (1968): Flood inundation and effects of urbanisation in metropolitan Charlotte, N.C., *Geol. Surv. Water Supply Pap.* 1591 C.

de Martonne, E. (1948): *Traité de géography physique*, Vol. II, 8th edn., A. Collins, Paris.

McLaughlin, T. G. (1947): Accelerated channel erosion in the Cimmaron Valley in Southwestern Kansas, *Journ. Geol.* **55**, 76–93.

McPherson, M. B. (1969): The nature of changes in urban watersheds and their importance in the decades ahead. In W. L. Moore and C. W. Morgan (eds.), *Effects of Watershed Changes on Streamflow*, Univ. of Texas Press, pp. 157–164.

McWhirter, N. and R. McWhirter (1975): *Guinness Book of Records*, 22nd edn.

Mead, W. R. (1954): Ridge and furrow in Buckinghamshire, *Geogrl. Journ.* **120**, 34–42.

Meade, R. H. and S. W. Trimble (1974): Changes in sediment loads in rivers of the Atlantic drainage of the United States since 1900, *IAHS-AISH Spec. Publ.* 113, 99–104.

Mensching, H. (1951): Akkumulation und Erosion niedersächsischer Flüsse seit der Risszeit, *Erdkunde* **5**, 60–70.

Mining Chamber of South Africa, *Mining Survey*, Johannesburg (monthly review).

Ministry of Town and Country Planning (1948): *Report of the Advisory Committee on Sand and Gravel, Part XVIII: Final Report*, HMSO, London.

Ministry of Town and Country Planning (1953): *Report of the Advisory Committee on Sand and Gravel, Part XVIII. Final Report*, HMSO, London.

Mirtskhoulova, T. E. (1974): Prediction of erosion intensification by man, *IAHS-AISH Spec. Publ.* **113**, 78–82.

Miyata, M. (1969): Variation in runoff and flood damage caused by urbanisation in the Shakujii River basin, *Geogrl. Rev. of Japan* **42**, 667–680 (in Japanese with English summary).

Mohr, E. C. J. and F. A. Van Baren (1954): *Tropical Soils*, Interscience Publ., London.

Morris, F. B. and G. F. Pyle (1971): The social environment of Rio de Janeiro in 1960, *Econom. Geogr.* **47**, 286–299.

Mosley, M. P. (1975): Channel changes on the river Bollin, Cheshire, 1872–1973, *The East Midland Geogr.* **6**, part 4, 185–199.

Mulavz, S. and S. Rybicki (1977): Subgrade and dump deformation caused by dumping of mine waste, *Eng. Geol.* **11**, 189–200.

Murgill, A. W., W. B. McMullen, M. Baker and J. F. McPhail (1976): Design and construction of roads on muskeg in arctic and sub-arctic regions. In Proceed. 16th Muskeg Research Cong., October 1975, *Techn. Memorandum 116*, N.R.C. of Canada, pp. 158–185.

Murphy, F. C. (1958): Regulating flood plain development, *Dept. of Geography, Univ. of Chicago, Res. Pap. no. 56.*

National Environm. Research Council (1972): *The Sand and Gravel Resources of the County around Witham, Essex*, Rep. No. 72/6.

Neev, D. (1965): *The Dead Sea*, Ph.D. Thesis, Hebrew University, Jerusalem.

Nelson, J. G. (1966): Man and geomorphic processes in the Chemung River valley, New York and Pennsylvania, *A.A.A.G.* **56**, 24–32.

Nicholson, M. (1970): *The Environmental Revolution*, New York.

Nikolayenko, U. T. (1974): The role of forest stands in the control of erosion processes and other negative natural phenomena, *IAHS-AISH Spec. Publ.* **113**, 83–86.

Nir, D. (1958): Etude sur la morphologie littorale d'Israël, *Annls Géogr.* 424–436.

Nir, D. (1968): La Vallée de Beth-Chèane, la mise en valeur d'une région à la lisière du désert, *Ecole Pratique des Hautes Etudes, VI section, Etudes et Mémoires*, No. 65, Paris.

Nir, D. (1974): *The Semi-arid World: Man on the Fringe of the Desert*, Longman, London.

Nir, D. (1975): Géomorphologie d'Israël, *Service de Cartographie et Documentation Géographiques, Mémoires et Documents* **16**, C.N.R.S., Paris.

Nir, D. (1977): Man as an agent of geomorphological processes changing the relief of the Earth, *Proceed. Eighth Sci. Conf., Isr. Ecolog. Soc.*, Tel Aviv, 30–31 May 1977, pp. 296–305.

Nir, D. and M. Klein (1974): Gully erosion induced in land-use in a semi-arid terrain (Nahal Shikma, Israel), *ZfG. Suppl.* **21**, 191–201.

Nir, Y. (1971): *Report on Improvements and Defence of Coast Line in Western Europe*, Min. of Development, Geolog. Survey (Israel) (Hebrew, polycopied).

Orme, A. R. and R. G. Bailey (1970): The effect of vegetation conversion and flood discharge on stream channel geometry; the case of Southern California watersheds, *Proc. Ass. Amer. Geogr.* 101–106.

Oxenham, J. R. (1966): *Reclaiming Derelict Land*, London.

Palmer, A. G. (1972): Settlement of a pipeline on thawing permafrost, *Transp. Engineer. Journ. ASCE*, 98, Proc. Paper 9090, pp. 474–491.

Papasov, R. (1974): L'indice d'érosion—facteur de determination de la turbidité des rivières, *IAHS-AISH Spec. Publ.* **113**, 15–20.

Parizek, R. R. (1971): Impact of highways on the hydrogeological environment. In D. R. Coates (ed.), *Environmental Geomorphology*, State University of New York at Binghamton, pp. 151–199.

Park, C. C. (1976): *Man-induced Changes in Stream Channel Capacity*, Inst. Brit. Geogr., Convention of Coventry, January 1976 (abstract, polycopied).

Parsons, J. S. and W. A. Bowen (1966): Ancient ridged fields in the San Jorge River floodplain, Colombia, *Geogrl. R.* 318–343.

Parsons, J. S. and W. M. Denevan (1967): Pre-Columbian ridged fields, *Scientific Amer.* **217**, 92–100.

Partosedono, R. S. (1974): Effects of man's activities on erosion in rural environments and a feasibility study of rehabilitation, *IAHS-AISH Spec. Publ.* **113**, 53–56.

Paxton, J. (ed.): *The Statesman's Yearbook for the Year 1974/75*, London.

Pécsi, M. and F. Probald (eds.) (1974): *Man and Environment*, Budapest.

Pereira, H. C. (1977): Land use in semi-arid Southern Africa, *Phil. Trans. R. Soc., London, B* **278**, 555–563.

Perry, R. A. (1977): The evaluation and exploitation of semi-arid lands: Australian experience, *Phil. Trans. R. Soc., London, B* **278**, 493–505.

Poland, J. F. and G. H. Davis (1956): Subsidence of the land surface in the Tulare Wasco (Delano) and Los Banos-Kettleman City area, San Joaquim Valley, California, *Trans. Amer. Geophys. Union* **37**, 287–296.

Potts, E. L. J. (1975): Mining subsidence and the environment. In M. J. Jones (ed.), *Minerals and Environment*, Inst. of Min. and Metallur., pp. 661–683.

Powell, M. D., W. C. Winter and W. P. Bodwitch (1969): Effect of urban development on flood discharges; current knowledge and future trends, *Proceed. Amer. Soc. of Civil Engineers, 99, Paper 6355, Hydrog. Div.*, pp. 287–309.

Powell, M. D., W. C. Winter and W. P. Bodwitch (1970): *Urban Soil Erosion and Sediment Control; Final Report on Community Action for Urban and Suburban Soil Erosion and Sedimentation Control*—Grant 15030 DTL, Nat. Assoc. of Counties Research Foundation, U.S. Dept. of Interior, Washington, D.C.

Ramser, C. E. (1917): Prevention of the erosion of farm lands by terracing, U.S. Dept. of Agric., Bull. No. 512 (reprinted in D. R. Coates (ed.), *Environmental Geomorphology and Landscape Conservation*, Benchmark Papers in Geology **8** (1973), 299–313).

Rathjens, C. (1979): *Die Formung der Erdoberfläche unter dem Einfluss des Menschen*, Teubner, Stuttgart.

Reastad, E. (1975): Effects of reservoir spillways and outlets on flood discharges. In E. Hansen (ed.) (1975), *Nordic Intern. Hydrolog. Decade Report No. 8*, pp. 74–79.

Reichman, S. (1976): Transportation infrastructure in adverse environmental conditions: the Elat-Ofira road, *Studies in Geography of Israel* **9**, 102–113 (Hebrew).

Reifenberg, A. (1950/51): Caesarea, a study in the decline of a town, *Isr. Expl. Jour.* **1**, 20–32.

Report of the extraction of Zifzif (1946): Ministry of Interior (Israel), polycopied (Hebrew).

Report on the Tribunal appointed to inquire into the disaster of Aberfar on October 21st, 1966 (1967): HMSO, London.

Richards, K. S. and R. Wood (1976): *Water Redistribution and Channel Process*, Inst. Brit. Geogr. Convention, Coventry, January 1976 (abstract, polycopied).

Rosenbery, J. L. and W. D. Klimstra (1965): Recreational activities on Illinois strip-mined lands, *Journ. of Soil and Water Conserv.* **19**, 107–110.

Rosepiler, M. J. and R. Reilinger (1977): Land subsidence due to water withdrawal in the vicinity of Pecos, Texas, *Eng. Geology* **11**, 295–304.

Royal Commission on Coast Erosion and Afforestation (1911): *Third (and Final) Report of the Royal Commission Appointed to Inquire into and Report on Certain Questions Affecting Coast Erosion, the Reclamation of Tidal Lands and Afforestation in the U.K.*, Vol. 2, Part 1, HMSO, London.

Ruhe, R. U. (1971): Stream regimen and man's manipulation. In D. R. Coates (ed.), *Environmental Geomorphology*, State University of New York at Binghamton, pp. 9–23.

de Salis, H. R. (1904): *Brandshaw's Canals and Navigable Rivers of England and Wales. A Handbook of Inland Navigation for Manufactures, Merchants, Traders and Others*, London.

Salmon, J. H. M. (1973): *A History of Goldmining in New Zealand*, Wellington.

Sauer, C. O. (1931): Cultural Geography, in *Encyclopaedia of the Social Sciences*, VI, pp. 621–623, Macmillan, New York.

Savini, J. and J. C. Kammerer (1961): Urban growth and the water regimen, *U.S. Geol. Surv. Water Supply Pap.* 1591-A.

Sawyer, R. M. (1963): Effect of urbanisation on storm discharges and ground water recharge in Nassau County, New York, *U.S. Geol. Surv. Prof. Pap.* **475-C**, 185–187.

Scheidt, M. E. (1971): Environmental effects of highways. In T. R. Detwyler (ed.), *Man's Impact on Environment*, New York, pp. 419–427.

Schick, A. P. and M. Inbar (1972): Some effects of man on the hydrological cycle of the Upper-Jordan–Lake Kinneret watershed, 2nd report of the ICU—Commission on the Intern. Hydrogr. Decade. Run-off Regimen and Water Balance II, *Freiburger Geogr. Hefte* **12**, 27–43.

Schumm, S. A. (1956): Evolution of drainage systems and slopes in badlands of Perth Amboy, New Jersey, *Geol. Soc. Amer. Bull.* **67**, 597–646.

Schumm, S. A. (1963): Sinuosity of alluvial rivers on the Great Plains, *Geol. Soc. Amer. Bull.* **74**, 1089–1100.

Schumm, S. A. and R. W. Lichty (1961): Recent floodplain formation along the Cimarron River in Kansas, *U.S. Geol. Surv. Prof. Pap.* **424-B**, 112–114.

Scott, K. M. (1973): Scour and fill in Tujunga Wash, a fanhead valley in urban Southern California, *U.S. Geol. Surv. Prof. Pap.* **721-B**.

Seaburn, G. E. (1969): Effects of urban development on direct runoff to East Meadoe Brook, Nassau County, Long Island, NY, *U.S. Geol. Surv. Prof. Pap.* **607-B**.

Seginer, I. (1966): Gully development and sediment yield, *Journ. of Hydrolog.*, 236–253.

Selby, M. J. (1972): The relationship between land use and erosion in the Central North Island, *N.Z. Journ. of Hydrolog.* **11**, 73–87.

Sharp, A. L., J. J. Bond, J. W. Neuberger, A. R. Kuhlman and J. K. Lewis (1964): Runoff as affected by intensity of grazing on rangeland, *Journ. of Soil and Water Conserv.* **19**, 103–106.

Sherlock, R. L. (1922): *Man as a Geological Agent; an Account of his Action on Inanimate Nature*, London.

Sherlock, R. L. (1931): *Man's Influence on the Earth*, London.

Simons, M. (1967): *Deserts—the Problem of Water in Arid Lands*, London.

Singh, T., Y. P. Kalra and G. R. Hillman (1974): Effects of pulpwood harvesting on the quality of stream water of forest catchments representing a large area in Western Alberta, Canada, *IAHS-AISH Spec. Publ.* **113**, 21–27.

Sion, J. (1940): Les raisons du dessin du cadastre dans la région Montpellieraine, *Bull. Soc. Languedoc de la Géogr.*, Montpellier, 19–28.

Slaymaker, O. and R. E. Gilbert (1972): Geomorphic process and land use changes in the coast mountains of British Columbia, *Les Congrès et Colloques de l'Université de Liège*, Vol. 67, Sympos. Intern. de Géomorph., pp. 268–279.

Smith, D., C. P. Twidale and J. A. Bourne (1975): Kappakoola dunes—aeolian landforms induced by man, *Austr. Geogr.* **13**, 90–96.

Smith, D. D. and W. H. Wischmeier (1962): Rainfall erosion, *Advances in Agronomy* **14**, 109–148.

Smith, D. I. and E. M. Bridges (1965): Report of a symposium (Rates of erosion and weathering in the British Isles) led at Bristol, January 1965, *Inst. of Brit. Geogr. Transact.*, No. 36, pp. XIII–XIV.

Smith, R. T. (1975): Early agriculture and soil degradation. In J. G. Evans, S. Limbrey and H. Cleere (eds.), *Effects of Man on the Landscape*, Res. Pap. No. 11, Counc. for British Archaeology, pp. 27–37.

So, C. L. (1971): Mass movement associated with the rainstorm of June 1966 in Hong-Kong, *Inst. of Brit. Geogr. Transact.* **53**, 55–65.

Soliman, M. M. (1974): Urbanisation and the processes of erosion and sedimentation in the river Nile, *IAHS-AISH Spec. Publ.* **113**, 123–126.

Sparrow, S. D., F. J. Wooding and E. H. Whiting (1978): Effects of off-road vehicle traffic on soils and vegetation in the Denali Highway region of Alaska, *Journ. of Soil and Water Conservation* **31**, 20–27.

Spencer, J. E. and G. A. Hale (1961): The origin, nature and distribution of agricultural terracing, *Pacific Viewpoint* **2**, 1–40.

Stall, J. B. (1966): Man's role in affecting the sedimentation of streams and reservoirs, Proceed. 2nd Amer. Water resour. Conf. (reprinted in D. R. Coates (ed.)), *Environmental Geomorphology and Landscape Conservation*, Benchmark Papers in Geology **8** (1973), 103–119.

Steinbrenner, E. C. and S. P. Gessel (1955): The effect of tractor logging on physical properties of some forest soils in SW Washington, *Proceed. Soil. Sci. Soc. Amer.* **19**, 372–376.

Stewart, O. C. (1956): Fire as the first great force employed by man. In W. L. Thomas (ed.), *Man's Role in Changing the Face of the Earth*, University of Chicago Press, Chicago, pp. 115–133.

Stockdale, F. A. (1928): Prevention of soil erosion during re-afforestation in hilly sections of the wet tropics, *Empire Forest. Journ.* **7**, 41–45.

Stocking, M. A. (1972): Aspects of the role of man in erosion in Rhodesia, *Zambezia*, **2**, 1–10.

Strahler, A. N. (1956): The nature of induced erosion and aggradation. In W. L. Thomas (ed.), *Man's Role in Changing the Face of the Earth*, University of Chicago Press, Chicago, pp. 621–638.

Strahler, A. N. and A. H. Strahler (1973): *Environmental Geoscience: Interactions between Natural Systems and Man*, Santa Barbara.

Sylvester, R. O. and F. B. De Waller (1972): Character and significance of highway runoff waters, *Dept. of Civil Eng., University of Washington, Seatle, Res. Pap.* Y-1441.

Talbot, W. J. (1961): Land utilisation in the arid regions of Southern Africa, *Arid Zone Research* **17**, 299–331.

Taylor, C. (1973): *The Cambridgeshire Landscape*, London.

Teller, H. L. (1968): Impact of forest land use on floods, *Unasylva* (FAO, Rome) **22**, 18–20.

Thomas, T. M. (1956): Wales: land and mines and quarries, *Geogrl. R.* **46**, 58–81.

Thomas, W. C. (ed.) (1956): *Man's Role in Changing the Face of the Earth*, University of Chicago Press, Chicago.

Thompson, J. R. (1970): Soil erosion in the Detroit metropolitan area, *Journ. of Soil and Water Conserv.* **25**, 8–10.

Tixeront, J. (1974): La couverture végétale et les débits des cours d'eaux dans le bassin de la Seine, *IAHS-AISH Spec. Publ.* **113**, 28–35.

Tóth, A. and Z. Fekete (1974): The effect of agriculture on the surroundings of Lake Balaton, *IAHS-AISH Spec. Publ.* **113**, 36–39.

Tricart, J. (1953): La géomorphologie et les hommes, *RdGD* **4**, 113–156.

Trimble, S. W. (1970): The Alcovy river swamp, the result of culturally accelerated sedimentation, *Bull. Georgia Acad. of Sci.* **28**, 131–141.

Turner, A. R. (1955): How Saskatchewan dealt with her "dust bowl", *Geogrl. Magaz.*, 182–192.

Unesco Statistical Yearbook, 1977, UNESCO, New York.

United Nations Food and Agricultural Organization (1977), New York.

United Nations Statistical Yearbook (1975): United Nations, New York.

United National Statistical Yearbook (1977): United Nations, New York.

United Nations Statistical Yearbook (1978): United Nations, New York.

U.S. Dept. of Agriculture (1968): Restoring surface-mined land, Misc. Publ. 1082 (reprinted in D. R. Coates (ed.)), *Environmental Geomorphology and Landscape Conservation*, Benchmark Papers in Geology **8** (1973), 414–430.

U.S. Dept. of Interior (1971): Impact of surface mining on environment. In T. R. Detwyler (ed.), *Man's Impact on Environment*, New York.

Verger, F. (1968): *Marais et Wadden du littoral français*, Bordeaux.

Vice, R. B., H. P. Guy and G. E. Ferguson (1969): Sediment movement in an area of sub-urban highway construction, Scott Run Basin, Fairfax county, Virginia, 1961–1964, *U.S. Geolog. Surv. Water Supply Pap.*, 1591-E.

Vita-Finzi, C. (1963): Carbon[14] dating of medieval alluvium in Lybia, *Nature* **198**, 880.

Vita-Finzi, C. (1964): Synchronous stream deposition throughout the Mediterranean area in historical times, *Nature* **202**, 1324.

Vita-Finzi, C. (1974): Physiographical effects of Man, *Enc. Brit.*, edition 1974, **14**, 429–433.

Vogt. J. (1953): Erosion des sols et technique de culture en climat tempére maritime de transition, *RdGD* **4**, 157–183.

Vosknil, W. H. (1930): *Minerals in Modern Industry*, New York.

Waananen, A. C. (1969): Urban effects on water yield. In W. L. Moore and C. W. Morgan (eds.), *Effects of Watershed Changes on Streamflow*, Univ. of Texas Press, pp. 169–182.

Wagret, P. (1968): *Polderland*, London.

Walling, D. E. (1974): Suspended sediment production and building activity in a small British basin, *IAHS-AISH Spec. Publ.* **113**, 137–144.

Warnock, R. G. and R. G. Lagoke (1974): Suspended sediments from urban development, *IAHS-AISH Spec. Publ.* **113**, 117–122.

Weitz, R. (ed.) (1973): *Urbanisation and the Developing Countries*, New York.

Whitaker, G. L. (1971): Changes in the elevation of Great Salt Lake caused by man's activity in the drainage basin, *U.S. Geol. Prof. Pap.* **750-D**, 187–189.

Whitfield, A. J. and J. A. Perrin (1939): Sand dune reclamation in the Southern Great Plains, U.S. Dept. of Agric., *Farmers' Bull.* No. 1825 (reprinted in D. R. Coates (ed.)), *Environmental Geomorphology and Landscape Conservation*, Benchmark Papers in Geology **8** (1973), 349–361.

Williams E. L. (1974): African giant—the saga of Cabora Bassa, *Optima* 95–106.

Williams, G. P. and H. P. Guy (1971): Debris avalanches—a geomorphological hazard. In D. R. Coates (ed.), *Environmental Geomorphology*, State University of New York at Binghamton, pp. 25–46.

Williams, R. J. (1974): Predicting sediment yield frequency for rural basins to determine man's effect on long-term sedimentation, *IAHS-AISH Spec. Publ.* **113**, 105–108.

Williams, W. W. (1960): *Coastal Change*, London.

Wilmoth, B. M. (1969): Environmental hazard of West Virginia, *Proceed. West Virg. Acad. of Sci.* **41**, 174–183.

Witthauer, K. (1975): Millionenstaedte um 1970/72, *Peterm. Geogr. Mitt.* **119**, 313–314.

Woeikof, A. I. (1901): De l'nfluence de l'homme sur la terre, *Annls. Géogr.* **10**, 97–114; 193–215.

Wolman, M. G. (1967): A cycle of sedimentation and erosion in urban river channels, *Geogrl. Annaler* **49 A**, 385–395.

Wolman, M. G. (1974): Fluvial processes, *Enc. Brit.* edition 1974, **7**, 437–446.

Wolman, M. G. and A. P. Schick (1967): Effects of construction on fluvial sediment, urban and suburban areas of Maryland, *Water Resources Res.* **3**, 451–464.

Wood, A. (1950): *The Groundnut Affair*, London.

Woolridge, S. W. and S. H. Beaver (1950): The working of sand and gravel in Britain: a problem in land-use, *Geogrl. Journ.* **115**, 42–57.

Yaalon, D. H. and B. Yaron (1966): Framework for man-made soil changes—an outline of metapedogenesis, *Soil Sci.* **102**, 272–277.

Yorke, T. H. and W. J. Davis (1971): Effects of urbanisation on sediment transport in Bel Pre Creek basin Maryland, *U.S. Geol. Surv. Prof. Pap.* **750 B**, 218–223.

Zapletal, L. (1960): Anthropogenous forms of relief, *Abst. of Papers, XIX I.G.U. Congress*, Stockholm, pp. 323–324.

Ziapkov, C. (1974): Economic activities influencing the fluvial regimen of the Pazardjik and Plovdiv plain. In M. Pécsi and F. Probald (eds.), *Man and Environment*, pp. 157–162.

Zube, E. (1966): A new technology for Taconite badlands, *Landscape Architecture* 136–140.

Index

GPSR Compliance

The European Union's (EU) General Product Safety Regulation (GPSR) is a set of rules that requires consumer products to be safe and our obligations to ensure this.

If you have any concerns about our products, you can contact us on ProductSafety@springernature.com

In case Publisher is established outside the EU, the EU authorized representative is:

Springer Nature Customer Service Center GmbH
Europaplatz 3
69115 Heidelberg, Germany

Batch number: 10607214

Printed by Printforce, the Netherlands